PROVING A MIRACLE

Proving a Miracle

Unlocking the Power of Prayer to Heal

Joshua W. Brown, PhD

HARPER

An Imprint of HarperCollins*Publishers*

This book contains advice and information relating to health care. It should be used to supplement rather than replace the advice of your doctor or another trained health professional. If you know or suspect you have a health problem, it is recommended that you seek your physician's advice before embarking on any medical program or treatment. All efforts have been made to assure the accuracy of the information contained in this book as of the date of publication. This publisher and the author disclaim liability for any medical outcomes that may occur as a result of applying the methods suggested in this book.

Names and identifying characteristics of some individuals have been changed to protect their privacy.

 For information, address HarperCollins Publishers, 195 Broadway, New York, NY 10007. In Europe, HarperCollins Publishers, Macken House, 39/40 Mayor Street Upper, Dublin 1, D01 C9W8, Ireland.

hc.com

HarperCollins books may be purchased for educational, business, or sales promotional use. For information, please email the Special Markets Department at SPsales@harpercollins.com.

FIRST EDITION

Designed by Bonni Leon-Berman

Library of Congress Cataloging-in-Publication Data has been applied for.

ISBN 978-0-06-342044-1

26 27 28 29 30 LBC 5 4 3 2 1

For the believers who doubt and
the doubters who believe

To my family

In memory of Andrew Snekvik

Though experience be our only guide in reasoning concerning matters of fact; it must be acknowledged, that this guide is not altogether infallible, but in some cases is apt to lead us into errors. . . . A wise man, therefore, proportions his belief to the evidence.

—David Hume,
An Enquiry Concerning Human Understanding

CONTENTS

INTRODUCTION

A Neuroscientist's Unplanned Search for Healing

FEBRUARY 2004. A COLD, WINTRY DAY IN THE SUBURBS OF St. Louis. I am thirty years old: a husband, a new father, and a neuroscientist awaiting publication of my first paper in *Science* magazine. I'm also a patient—a man about to retrieve the results of his own latest magnetic resonance imaging (MRI) test. Six months ago, I had a seizure. Shortly after that, I was diagnosed with a glioma, a tumor that begins in the glial cells of the brain or the spinal cord. Gliomas can be slow-growing at first, but they are deadly, and deterioration is inevitable. It is only a question of when the decline will begin—and how much longer I have left.

Throughout my career, I've relied on MRIs to gain insight into the brain's inner workings, but in mere moments it will be images of my own brain I'll be studying. My heart is racing, and I feel a lump in my throat as I stand at the records window in the bland hospital hallway of the radiology lab. I hope against hope that my tumor has not advanced. The usual course for a glioma entails increasingly severe headaches, gradual distortions in vision, and possible personality shifts that will cause me to cease caring about everything that matters to me right now. Sooner or later, science tells me, I will face this descent.

But I also possess the seeds of faith that things might not be as dire as they seem and that I might be spared the worst. It's a conviction that few in my inner circle share, but lately, since the diagnosis, I've grown increasingly aware of the possibility of healing through prayer, thanks to a community of believers at my church. I haven't been one to believe in miracles, but now, with my own health at stake, I wonder if a shred of belief in the otherwise unlikely might not be such a bad thing.

I am not yet experiencing any of the dire symptoms of decline suggested by conventional science. Can I be healed by prayer?

Proving a Miracle is my story. It also explores the ways in which our beliefs, and our spiritual beliefs in particular, might affect our health. I begin by asking basic questions about why scientists have been reluctant to study spiritual beliefs and what we now know, scientifically, about the impact of the spiritual on our physical selves. How do we form and maintain beliefs from a neuroscience perspective? And how do those beliefs in turn affect our health? How do we adjust or update our beliefs, especially as we seek out new experiences or when new information comes to light?

Subsequently, I consider questions about how the beliefs that guide us might heal us. How do social practices that include the fostering of community, forgiveness, and gratitude transform our brains? Does healing prayer work, and if so, what can studying placebos and other mind-body effects tell us about how prayer heals? How might cultivating a lifestyle of hope, resilience, learning, and spiritual growth affect our mental and physical health?

Finally, I look at efforts to push scientific study of the effects of spiritual beliefs into uncharted territory. How can we empirically examine events—such as healings ascribed to miracles—that may traditionally have been viewed as beyond science's reach without bolstering unjustified supernatural claims?

The field of neuroscience has seen tremendous growth over the past three decades. There aren't just new technologies and capabilities; there's new understanding about how our thoughts and beliefs alter our biology. Scientific studies have led to widespread acceptance of secular meditation's health benefits, such as better immune response and positive mental states. Yet skepticism still reigns among scientists when it comes to studying the benefits of spiritual beliefs. According to a 2024 Pew Research Center study, 79 percent of US adults believe there is "something spiritual beyond the natural world, even if we cannot see it." Besides the 58 percent of "religious" Americans who participate in organized activities of a church, synagogue, or mosque, another 21 percent of the population express "spiritual" beliefs. One common spiritual belief is that prayer to God or a higher power can result in miraculous healing. A reported 73 percent of US medical doctors believe that miraculous—defined as medically inexplicable—recoveries occur today. Fully 27 percent of US adults claim to have *personally* "experienced a physical healing that could only be explained as a miraculous healing." In dozens of countries, those numbers are even higher. Why refuse to examine how spiritual beliefs that so many people find helpful affect our brains and bodies? The scientific method was designed for questions such as these.

At the Global Medical Research Institute (GMRI), which I cofounded in 2011, we are testing the empirical effects of prayer on healing. I share our results in the following pages. I'll also share the work of others who are exploring how spiritual beliefs provide social support and facilitate pain and symptom relief. There are still unanswered questions regarding the exact mechanisms of relief and cure—investigations into which could alleviate human suffering.

My life and my work inspire each other. As a neuroscientist and a professor of psychological and brain sciences, I bring a healthy amount of scientifically based skepticism to the table. I'm also a Christian who experienced a remarkable medical recovery. I've always lived in the two worlds of science and faith. A skeptic. A believer.

My personal experience with healing prayer led me to push my scientific research into new frontiers of empirical inquiry. I've seen people of faith prematurely conclude that God has performed a miracle when there is a simpler natural explanation. I've also seen neuroscientists hesitate to investigate how spiritual beliefs interact with the brain and body for fear that they will be greeted by skepticism, scorn, or worse. Neither point of view has convinced me. Instead, it has made me question if religious adherents forfeit health-promoting insights made available by scientific research, while scientific naturalists miss opportunities to learn about other factors, including spiritual beliefs, that can and do promote healing.

The question for science is not whether a supernatural being or metaphysical force *causes* healing effects. Rather, as scientific researchers develop improved tools to examine *empirical effects* of spiritual beliefs, the question of what those practices and effects are has become a feasible area of study. We may not be able to strip the language of belief out of prayer in the way that it's practiced, but we can study the empirical effects of prayer while remaining agnostic about whether there's a supernatural or metaphysical mechanism at work.

One might ask whether we should study the effects of beliefs across all religious and spiritual traditions. My experience is with Christianity. I have journeyed to places all over the world, from Cuba to Mozambique, where Christians claim to have experienced miraculous healing through prayer. While my current research focuses on Christian beliefs, I support the study of other beliefs and am intrigued by the possibility that diverse religious and spiritual beliefs have different effects on the brain, its activity, and our health. Asking whether a focus on the breath relieves anxiety, such as we see in Buddhist-based experiments, is different from asking whether praying the rosary relieves anxiety—but that's the point. We won't know how the specific contents of spiritual beliefs affect the mind and body until we study them.

People also ask me if healing prayer is effective, or why some people are healed and not others. I wish more definitive answers were available to us, but this book explores that mystery alongside the science and experiential evidence that is available.

Shortly after I experienced healing, a close friend died from colon cancer while in his mid-thirties, leaving behind a widow and three young children. I can make no guarantees. My goal in sharing my journey is to communicate what I've learned about spiritual beliefs with the hope and intention of helping others. In my travels, I've witnessed many inspiring cases of healing. No matter the situation, or how dire, I have found hope. I'd like to pass that hope on to those who are suffering, to those in pain, and to those seeking to be healed.

Twenty-two years ago, I set off on an unplanned, unwanted, and traumatic search to find healing for myself and to explore the intersection of science and faith. I discovered much that surprised me and changed my life. I invite you to follow along as I present the evidence of how spiritual beliefs can help us to experience healthier minds and bodies and a deeper sense of well-being.

PART I

The Science of Belief

1

Should Scientists Study Spiritual Beliefs?

THE DAY BEFORE MY LIFE CHANGED WAS A PERFECT SUMMER day. My wife, Candy, and I were eagerly awaiting the birth of our first child. Impatient for and exhilarated at what was to come, I spent the afternoon bicycling with friends around the redbrick suburbs of St. Louis. We pushed each other on the climbs, flying headlong into the warm, humid winds on the descent and soaking up the last few minutes of pure freedom. That night, exhausted, I fell asleep quickly. The next thing I recall is jolting awake around five the next morning. It was August 2, 2003, and I was lying in the back of an emergency vehicle, with no idea how I had gotten there. In the sterile, brightly lit ambulance bay, I had an oxygen mask over my nose and mouth, an IV in my arm, and paramedics swarming around me. Candy, heavily pregnant and uncomfortable, was there beside me wearing an expression that was both comforting and concerning. I later learned she'd awoken half an hour earlier, looked over, and saw me lying on my back staring straight up at the ceiling, neither awake nor asleep. She tried to rouse me, but I was unresponsive. My eyes were wide open, my pupils fixed and dilated. When I let out a piercing shriek and my right arm shot straight up into the air, Candy ran to the telephone to call 911. Shortly thereafter, I began foaming at the mouth.

Soon I was in an ambulance, on my way to the hospital, where, among other tests, the doctors did a computerized tomography (CT) scan of my head. They wanted to rule out a hemorrhagic stroke, a brain bleed due to the rupture of a blood vessel that can cause sudden death. Thankfully, they found no sign of any kind of stroke. By late morning, the emergency room discharged me. The next day, I met with the neurologist that the ER doctors recommended. I wasn't particularly anxious about the visit. Mostly, I just didn't want to deal with a potential medical issue of my own when we had so much more to do to get ready for the baby. The neurologist ordered an MRI and an EEG (an electroencephalogram, a test that measures the electrical activity of the brain), two standard tests following a seizure.

Four days later, at the edge of dawn, I stood in the maternity ward, beaming with pride, bulky camcorder in hand, filming my newborn daughter, Katrina's, first bath. Then, while Candy was still in the hospital recovering from her C-section, I underwent my first MRI scan. We were home with our new baby when I got a phone call from the neurologist's office asking me to undergo a second MRI. I understood from my own studies with MRI as a neuroscientist that noise or unwanted interference is often introduced into such images. Power spikes in the machine or tiny pieces of metal in the subject's clothing can be the culprits—the reason that initial images can be at least partially unreadable. Though the functional neuroimaging equipment we used in the lab is even more sensitive to noise than standard hospital equipment, it seemed logical that image quality might be a problem for the radiologist too. I reasoned that I was simply following standard protocol.

But at the radiologist's office, the technician explained that this MRI was going to be a tumor protocol scan, a specific set of scans taken when a tumor is suspected. The neurologist's office hadn't said anything in our phone call about a tumor. I immediately shifted to high alert.

By the time I'd emerged from the second MRI, I'd already decided: Waiting another week for an appointment with the neurologist to get the results was out of the question. Something was going on in my brain, and I wanted to know what it was. The Health Insurance Portability and Accountability Act (HIPAA), passed in 1996, required the hospital to provide requested medical records. Given my background in neuroscience, I could reliably read the scans and the radiologist's report myself.

And so, the next day, I returned to the radiologist's office to retrieve the oversize brown manila envelope containing the films and readings. Nervous, I made my way back to my car—a red Honda Civic starter—opened the door, and slid in. Whatever I was going to learn, I wanted to learn in private. Unfastening the metal clasp sealing the envelope shut, I slipped the papers free and scanned to the bottom of the first page, where I saw the words "swollen prominence of the left parahippocampal gyrus. Appearance is worrisome for low grade tumor such as ganglioglioma or astrocytoma." A low-grade glioma is a tumor with a median survival rate of 5.9 years to 16.7 years, even with treatment.

I held the films up to the car window for additional light and could clearly see what the radiologist had seen—a bright enlarged area toward the lower middle area of the left side of my brain. The tumor was deep enough that it would be difficult to access surgically; even so, surgery might not prolong life with this type of tumor.

I had just read my death sentence at thirty years old. And though I had no idea what the time frame would be, I knew what science said came next—all the terrifying secondary effects of a brain tumor. I wish I could say my reaction was my usual calm, measured, collected manner. It wasn't. I was upset and afraid, worried about my wife and baby daughter's future.

When I arrived home from the radiologist's office, Candy and I prayed together. Candy had already taken on the role she would play throughout this saga: that of a supportive partner, the person I could always rationally talk to about my options, and a pillar of incomparable strength. During the subsequent sleepless nights, I tried to conjure up ways I could stave off my inevitable decline. A friend from Harvard, a cardiologist at the same St. Louis hospital where I received care, gave me access to UpToDate, an online service that informs medical professionals about the latest medical findings. As a neuroscience researcher, I got to work, gathering every detail I could on gliomas, their treatments, and their prognoses. Candy and I consulted with our pastor, and my parents and I had many long talks. My father is an ordained minister, who, back then, served as chaplain for a prison in the San Diego area. He is well practiced in offering comfort to men in trouble.

The immediate aftermath of my diagnosis was a whirlwind of prayer meetings, newborn care, and trying to maintain some facsimile of normal life. But even the good days felt darkened by the reveal of my tumor. In a video Candy made during that time, I'm shaking a rattle at my infant daughter and whirling her around to her great delight. It looks like any father and daughter deep in their joyful first bonding, but there was a strange sense that we were recording it not only to capture a joyful early childhood memory in the moment but also because we knew that there might not be many memories to create.

In September, I had the EEG test my neurologist recommended. The EEG was to help us pinpoint the location of the seizure generators so that they could be surgically removed. Unfortunately, the results were inconclusive. There was no clear map for where to operate. The next step was to see the neurosurgeon to consult about a possible biopsy. Excising the tumor *might* extend my life, and it *might* resolve the epilepsy (i.e., the seizures). But it wouldn't be curative. With gliomas, the cells diffuse too much into other parts of the brain for the surgery to eliminate the tumor or prolong life by much.

There was no real advantage in having the surgery right away given the risks of brain damage and low probability of cure. Palliative surgery would still be an option at a later stage, when the tumor got worse and we were looking to alleviate my symptoms, including the possibility of any ongoing seizures. In the meantime, my neurologist recommended continuing to monitor the tumor with quarterly MRIs. Regular scans, she told me, would allow her to keep an eye on whether it was growing, and whether it was transforming from a low-grade tumor to a high-grade tumor.

Which is how I found myself on that wintry day in February 2004 that I described in the introduction, waiting for the results of yet another MRI. I was emotionally in a very different place than when I was first diagnosed. As time passed, and the horror and shock subsided, a new question arose. The assumption that I was going to die was just that—an assumption. What if, by some miracle, there was a path to continue living?

And what if that path involved faith? By now, I'd met many individuals who not only believed that prayer *could* heal but had personally experienced healing they attributed to prayer. The prayer I'd been receiving had positively affected my sense of well-being. Perhaps there was something worth exploring here, and just maybe, applying my neuroscience training to study spiritual beliefs could help me—and others—to heal.

After all, *a diagnosis is not a prognosis, and a prognosis is just a prediction.* In the gap between the two, I reached for a sliver of hope. And I began asking larger questions about how spiritual beliefs form and affect health, and might even impact healing.

Why Scientists Distrust Spiritual Beliefs

The study of how spiritual beliefs work in the brain is not a widely embraced line of inquiry among scientists, even as the scientific study of beliefs more broadly has blossomed over the past half century. In the late 1970s through the 1990s, new functional brain imaging methods, including positron emission tomography (PET) and functional magnetic resonance imaging (fMRI), were developed. These advances made it possible to study the activity of the entire human brain at once, rather than a few cells at a time, making it easier to locate which parts of the brain are responsible for specific cognitive processes, including such basics as perception and memory, and higher processes like reasoning and creativity.

In the 1930s, famed American-Canadian neurosurgeon Wilder Penfield, seeking a treatment for epilepsy, located specific regions of the brain and created maps, millimeter by millimeter, of the areas responsible for sensory and motor processes. Using his novel Montreal procedure (named for its city of origin), Penfield put patients under a local anesthetic, cut into the skull to expose their brain tissue, and used electrical probes to stimulate their brains. He then asked the still-conscious patients what they felt. Their responses enabled him to pinpoint where the epileptic seizure activity was occurring and thus the tissue that he needed to excise to treat the epilepsy. They also enabled Penfield to build maps of specific regions of the brain.

This innovation advanced our understanding of basic cognitive processes. It required, however, an invasive surgical procedure that could only be performed ethically to treat epilepsy. Mapping one small area of the brain at a time was, moreover, impractical, given the remarkable range of configurations generated by the 100 billion brain cells that can be activated during our cognitive processes. PET and fMRI efficiently and noninvasively solved both problems, allowing activity in each part of the brain to be measured simultaneously and continuously.

Initially, the new neuroimaging research made possible by PET and fMRI focused on basic sensory processes like perception: how the brain processes what the eyes see. Over time, neuroscientists began to study higher cognitive functions, including beliefs. Elizabeth Phelps is a professor of human neuroscience at Harvard University. In the early 2000s, she and her research group used fMRI for a study of social cognition: how we process, store, and apply information in social situations. Specifically, she used neuroimaging techniques to identify activity in the *amygdala*, the almond-shaped part of our brain that processes fear, to investigate how the brain responds when we see someone of a different ethnicity. By the 2010s, the field of social cognitive neuroscience had become mainstream. Social relationships and beliefs could now be studied in terms of neurophysiological processes.

Social scientists have long investigated how our beliefs influence everything from our psychology, politics, sociology, judgment, and decision-making to our health outcomes. The emerging methods and technologies now enable natural scientists to explore belief formation, maintenance, and the effects of beliefs. Yet, even as the neural basis of belief became a hot topic in scientific research, investigations of the effects of spiritual belief did not become mainstream. To quote a colleague, "Neuroscientists have tended to steer clear of studying how people's [spiritual] beliefs affect their brains and vice versa."

Reluctance to investigate the neural basis of spiritual beliefs, even after technology was available to do so, reflects a long-standing conflict between defenders of science and religion. As far back as 1633, a Roman Catholic Church court convicted astronomer Galileo Galilei of heresy because he contradicted the Church's interpretation of the Bible as teaching that the sun moves around the earth. Charles Darwin's publication of *On the Origin of Species* in 1859 inflamed the conflict because his theory of evolution challenged how many Christians, Jews, and Muslims interpreted their scriptures' respective teachings about creation—namely, that God directly created life in its present forms.

In 1925, the evolution versus creation debate became a national spectacle here in the United States, as high-profile attorneys and reporters descended on the small town of Dayton, Tennessee, for what became known as the Scopes Monkey Trial. The Tennessee legislature had just passed the Butler Act, which forbade public school teachers from promoting "any theory that denies the story of the Divine Creation of man as taught in the Bible, and to teach instead that man has descended from a lower order of animals." The court found the defendant, John T. Scopes, guilty for teaching evolution and imposed a token fine, a ruling overturned on appeal. The state legislature repealed the Butler Act in 1967, and a year later, the US Supreme Court held similar laws unconstitutional in *Epperson v. Arkansas*. But the damage had been done. There was now a baseline distrust among scientists toward religious defenders of creationism as blocking scientific knowledge.

Dismayed by a pitched culture war, Harvard evolutionary biologist and self-described Jewish agnostic Stephen Jay Gould attempted to broker a peace. Gould empathized with Harvard students who sensed a conflict between being good Jews or Christians and being good scientists. In a 1997 essay, Gould introduced his *nonoverlapping magisteria* (NOMA) thesis. He assigned both science and religion their own magisteria, or domains of authority, that do not overlap. According to Gould, science's net extended over "the empirical universe, what it is made of (fact), and why it works this way (theory)" and religion's encompassed questions of "moral meaning and value." In Gould's view, science and religion address totally separate realities and exert equal authority over their distinct domains.

Perhaps unsurprisingly, given the heatedness of the science versus religion debate, conflict continued beyond Gould's peacemaking efforts. Reacting against Gould's NOMA thesis, Oxford biologist Richard Dawkins saw religious belief as antithetical to progress and human flourishing rather than merely a benign domain of authority separate from science. He spoke of a "God delusion" and the fundamental irrationality of belief in any higher power in a scholarly paper and his later book of the same name. Evolutionary biologist Richard Lewontin at Harvard worried further that ceding any space to religion would be a step backward for science in "the real struggle between science and the supernatural . . . because we have a prior commitment, a commitment to materialism . . . that materialism is absolute . . . we cannot allow a Divine Foot in the door." Entertaining the possibility of *anything spiritual* beyond the material world, Lewontin feared, risked a religious takeover. Skeptics like Michael Shermer (founding publisher of *Skeptic* magazine) accepted NOMA but rejected religious beliefs personally and, generally, as inimical to science. "I don't think there is a God, or any sort of anthropomorphic being who needs to be worshipped, who listens to prayers, who keeps a moral scoreboard that will be settled in the end, or who cares one iota about who wins the Super Bowl. There is no afterlife. We just die, and that's it," Shermer wrote in his essay "Why I Am an Atheist."

Apart from any threat they may pose to evolution or materialism, certain spiritual beliefs do not obviously lend themselves to scientific study. Faith claims such as Jesus of Nazareth's performance of miracles and resurrection from the dead are based on historical events. Because a historical event is rarely repeatable under the same conditions, it cannot be proven false—or true—using the scientific method of repeating a set of conditions in order to produce the same outcome. There are scientific methods like radiocarbon dating that use evidence to support historical claims such as that the Norse people reached North America (Newfoundland) around 1000 CE. But such methods generally draw on the principle of repeatable measurements. If one duplicates a radiocarbon-dating test on the same archeological artifact, one can expect to get the same date result, even though it is not possible to repeat the Norse migration of 1000 CE itself.

While some beliefs cannot be falsified, other beliefs persist despite having been falsified by scientific methods. Science proceeds by testing hypotheses and rejecting those contradicted by empirical evidence. Consider the claim by flat-earthers that NASA manipulates its imagery of the spherical earth and that the earth is flat. If scientists have empirically tested a belief such as whether the earth is flat or round and falsified the flat claim through multiple tests and approaches, they will not embrace attempts to resurrect the question. For working scientists, continued pursuit of already disproven beliefs can be perceived as a devaluation of science.

Spiritual beliefs, such as the efficacy of praying for healing, that can be empirically tested and have shown promise, may nevertheless be rejected as an object of study given limited resources. A systematic review of intercessory prayer published in 2009 concluded that "the evidence presented so far is interesting enough to support further study. However, if resources were available for such a trial, we would probably use them elsewhere." Alzheimer's research is one such area where scientists compete for resources. Research into beta-amyloid plaques long dominated the investigation of causes of Alzheimer's disease. Disgruntled researchers who wanted to study a new line of inquiry—the potential viral causes of Alzheimer's—felt so pitted against the dominant scientists that protesters labeled their opponents the "Amyloid mafia." The viral researchers were not wrong about the promise of their idea: A recent study demonstrated that vaccination against herpes zoster (shingles) also has a protective effect against dementia, suggesting that viral infections *are* an important contributor to the causes of Alzheimer's.

Science, as philosopher of science Thomas Kuhn once noted, is essentially a series of contests for *paradigm* dominance. Scientific paradigms are theoretical orientations that guide what types of questions, evidence, and methods researchers can use if they want to secure funding and respect. Paradigm shifts often involve intense competition—like that between the amyloid and viral approaches to Alzheimer's. Combine science's inherent competitiveness with the length of time it takes to reach definitive, practical results, and it's easy to see why scientists may prefer to shut the door to certain lines of inquiry, such as mechanisms of spiritual belief.

By 2005, the conflict between creationists and the supporters of evolution's teaching had been settled. In September of that year, a federal court tried the teaching not of evolution but of "intelligent design," a theory of the origins of life "that differs from Darwin's view." The Dover, Pennsylvania, school district required use of a biology textbook, *Of Pandas and People*, that cites the survival of cute but evolutionarily unfit pandas as evidence of an intelligent designer. The media dubbed *Kitzmiller v. Dover Area School District* the Dover Panda Trial, alluding to the Scopes Monkey Trial. The court ruled that intelligent design is not science. At issue was that intelligent design "invok[ed] and permit[ted] supernatural causation," thereby "abusing science" by "violating the centuries-old ground rules." The court determined that intelligent design "cannot uncouple itself from its creationist, and thus religious, antecedents." Evolution had won, and Stephen Jay Gould's NOMA peacekeeping became the rule of the day: Basic science research stayed on its side of the fence and religion stayed on its side, with very little meeting in the middle outside of the social sciences. Scientists who were also persons of faith were expected to avoid scientific studies of their faith, especially studies that might appear to promote religious or spiritual beliefs. Trespassing magisterial boundaries in this way could be deemed "abusing science."

Scientists Get Curious About Spiritual Beliefs and Their Effects

Evolution had secured a major victory, but another tectonic paradigm shift was underway in 2005, opening an unexpected front in the conflict between science and spiritual belief. Significantly, this shift took place directly at the intersection of mind, belief, practice, and the brain. It began with seminal studies on meditation by Richard J. Davidson, a professor of psychology at the University of Wisconsin–Madison, and Jon Kabat-Zinn, a professor of medicine at the University of Massachusetts. Kabat-Zinn's fascination with Buddhist meditation ignited in the 1960s when he was a student at MIT. His curiosity piqued by a campus flyer, Kabat-Zinn attended a talk by Philip Kapleau, a famous teacher of Zen. When Kapleau demonstrated the power of paying attention as if it really matters, it was a light-bulb moment for Kabat-Zinn. He became deeply curious about how living each moment to the fullest might change the quality of his life.

Kabat-Zinn and Davidson met in 1972, when Davidson was a grad student at Harvard and Kabat-Zinn had just received his PhD from MIT. Kabat-Zinn, who was trying to figure out what he wanted to do with his life, taught Davidson mindfulness meditation on Davidson's living room floor in Cambridge, Massachusetts. After his second year of graduate school, Davidson went on his first meditation retreat. He spent three months in India and Sri Lanka and returned home a true believer. Davidson saw through the lens of his personal experience that meditation affected the brain—and one's emotions. The meditators he knew were kind, warmhearted people, unlike some of his professors. Davidson fervently wanted to study the effects of meditation for his dissertation, but his professors made clear to him that he'd better switch topics if he wanted to succeed in the profession. So he did. He focused his studies on emotion and the brain instead.

A meeting with the Dalai Lama in 1992 would change everything. The Dalai Lama had learned of Davidson's work on emotion and invited him to visit Dharamsala, India, where the Dalai Lama resided. Ostensibly, the purpose of the visit was so that Davidson could interview Buddhist monks and learn more about how their meditation practices influenced their emotions. In the meeting, the Dalai Lama went a step further and asked why Davidson couldn't use the then-developing tools of neuroscience to study kindness and compassion, much as scientists were already studying the more negative emotions of pain and anxiety. It was the prompt Davidson needed. His old friend Kabat-Zinn had developed a program in meditation or mindfulness-based stress reduction (MBSR) for the Stress Reduction Clinic at the University of Massachusetts Medical Center. For MBSR, Kabat-Zinn combined practices based in Theravada and Mahayana Buddhism with yoga, and utilized crafted scripts that focused on how body awareness could reduce stress. Joining forces, Kabat-Zinn, still in Massachusetts, regularly flew out to Wisconsin, where Davidson worked at the time. They developed a randomized, controlled study of the effects of an eight-week MBSR program on the brain and immune system.

Davidson and Kabat-Zinn's subsequent 2003 paper, "Alterations in Brain and Immune Function Produced by Mindfulness Meditation," generated enthusiasm and emulation. The study measured electrical activity in the brains of healthy participants before and after the mindfulness program. The results showed that compared with nonmeditators, the meditators had a significant increase in left-sided anterior activation of the brain, a pattern previously associated with a positive emotional state over the long term.

The researchers also inoculated all the subjects—half of whom had completed the meditation course and half of whom hadn't yet started—with a flu vaccine. Next, they measured the subjects' immune response by looking at the level of antibodies in the blood. The result? The group that meditated had significantly higher antibody levels in response to the vaccine—essentially a much better immune response. Their meditation practice had both improved their emotional state and increased their ability to fight illness, a very strong finding that positive mood and mental states are associated with objectively better health.

This landmark study opened the door for others to test the biological effects of what were essentially spiritual practices. Not that any of the researchers responsible for the 2003 study described their work as spiritual. The MBSR practices used in these initial studies were framed as secular. In his quest to make mindfulness accessible to an American audience, Kabat-Zinn consciously stripped religious references, such as to Buddhadharma, or teachings of the Buddha. Kabat-Zinn focused instead on simple practices such as paying attention in the present moment and health benefits like stress reduction.

Similarly, in the early days of his research on the topic, Davidson emphasized emotion, his area of study in the field of psychology, not religion. Davidson was careful to state that he was investigating the biological effects of the new emotional states that meditation practices instilled. Ultimately, the Dalai Lama, with whom Davidson and Kabat-Zinn were publicly allied, gave a keynote speech to the Society for Neuroscience in 2005. In it, he followed NOMA rules: He told the assembled scientists that Buddhists and scientists could find common ground over "empirical observable facts of the human mind" or the effects of Buddhist practices, while "not falling into the temptation of reducing the framework of [Buddhism] into [the framework of science]," and vice versa.

The Dalai Lama, Kabat-Zinn, and Davidson might have believed in the early 2000s that they had avoided collapsing the two separate frameworks of religion and science. By denuding the meditation practices they were studying of spiritual language, they reported empirical findings on meditation's effects that did not rely on a particular spiritual framework.

Yet it was their shared interest in Buddhism that had brought the men together, and their mutual desire to research Buddhism and deepen their personal practices of it continued to bond them. Both Davidson and Kabat-Zinn served on the board of the Mind & Life Institute, an organization founded by Francisco Varela, a neuroscientist; Adam Engle, a lawyer; and the Dalai Lama. The Institute's mission statement was "to foster a creative dialogue between the Buddhist tradition and Western science," including the exploration of "new ideas and paradigms for collaborative research; to inform the Buddhist tradition of scientific findings relevant to its concerns; to *deepen scientific understanding of Buddhist thinking, and to contribute to its evolution in the West.*" Explicitly, they hoped to not only deepen scientific understanding of Buddhism but also contribute to the evolution of Buddhism.

The question of whether mindfulness can be effectively uncoupled from its religious roots remains contested. Candy Gunther Brown, a Harvard-trained professor of religious studies at Indiana University, has written numerous books on subjects ranging from evangelicalism to spiritual healing practices to the ethics of alternative medicine, including the ethics of mandatory mindfulness and yoga training in public schools. She is also my wife.

Brown explains that not all mindfulness is inherently Buddhist, but "nominally secular programs [in mindfulness] instill culturally and religiously specific and contested worldviews, epistemologies, and values." Kabat-Zinn understood mindfulness, in his own words, as the "essence of the Buddha's teachings," and he intended MBSR as a "launching platform" for "direct experience of the noumenous [*sic*], the sacred, the Tao, God, the divine"—all of which he readily admitted once MBSR gained scientific legitimacy. It's difficult to uncouple such an understanding of mindfulness from the Buddhist beliefs that underpin it.

In a recent book, *Altered Traits: Science Reveals How Meditation Changes Your Mind, Brain, and Body*, Davidson and another Buddhist-inspired collaborator, Daniel Goleman, acknowledge as much, discussing why effective mindfulness practice goes beyond five-minute daily scripts and requires "lightening our sense of self as the key to such inner freedom." Throughout *Altered Traits*, Davidson and Goleman refer to Buddhist concepts and describe how adventurous "Dan [Goleman] was as he hung out with yogis and lamas, and spent months poring over that fifth-century guidebook for meditators, the *Visuddhimagga*." They distinguish between fleeting "pleasant states" an individual may feel after practicing the MBSR version of mindfulness and more lasting, deeper, positive "altered traits" of practicing meditation as "highly seasoned practitioners in Asia" do.

As Davidson's, Goleman's, and Kabat-Zinn's work in mindfulness gained popular acclaim in the early 2000s, it intersected with another powerful paradigm shift: the neuroplasticity revolution. The basic premise of neuroplasticity is that the brain is literally rewired by experience. For example, if we learn to play the piano, the part of our brains that represents playing music expands. These empirical, observable changes aligned perfectly with what the mindfulness crowd was arguing: that mindfulness stimulated changes in mental and emotional states that could be studied as biological changes in the brain.

Davidson and Goleman's claim has always been that we change our brains as we change our beliefs, but by the time *Altered Traits* appeared in 2018, the beliefs they identified as changing our brains most were more overtly spiritual. The highest transformations occur, they now claimed, by looking through a specifically "Eastern lens." Though they never explicitly compared the transformations achieved by Buddhist meditation with possible transformations by Christian contemplative or intercessory prayer, they endorsed Buddhist practices as effecting "a radical transformation," the most brain-changing altered traits.

The net result? The wall between studying effects of spiritual beliefs and studying beliefs' biological effects on the brain crumbled. Buddhist meditation practices became subjects of scientific study partly because researchers felt they could extract and secularize specific meditation techniques, separating them from their traditional religious contexts. For the extractors, Buddhism seemed less intertwined with supernatural beliefs than other religious traditions and therefore more given to secular adaptations.

This assumption is not shared by all scholars of Buddhism, however. Distinguished Research Professor Ron Purser has argued that modern "McMindfulness" techniques are a distortion rather than an extraction of Buddhist practices. Professor of Buddhist Studies Rupert Gethin details how metaphysical and cosmological beliefs are "actually fundamental to Buddhist thought" throughout history since, for example, "we know of no Buddhism or Buddha that did not teach a belief in rebirth, or conceive of rebirth as fluid among different realms, whether animal, hellish, human, or heavenly." Davidson and Kabat-Zinn nevertheless maintained that individuals could practice mindfulness without adopting traditional Buddhist cosmology. In so doing, they helped define what scholar of Buddhism David McMahan calls Buddhist Modernism, a "new hybrid form of Buddhism that has emerged within the last 150 years . . . that includes liberal borrowing from scientific vocabulary . . . [that] reconfigures Buddhism as a kind of psychology." Modernizers like Davidson and Kabat-Zinn convinced a large segment of the population, including many fellow scientists, that secular mindfulness could be practiced and studied scientifically without violating NOMA's separation of science and religion.

Despite their attempts to downplay religious foundations of mindfulness, Davidson's and Kabat-Zinn's research set a precedent for scientific study of religious practices. Whether mindfulness can truly be secularized or not is a matter of ongoing debate, and I take no position on this question. Rather, the fact that this question remains contested means that there are still people who perceive scientific studies of mindfulness as scientific studies of de facto religious practices, not just psychological or biological phenomena. The cultural perception that Davidson and Kabat-Zinn used science to study religion helped undermine NOMA. Gould's idea of separate realities for science and religion was, in short, losing influence.

Davidson, Kabat-Zinn, and their colleagues aren't the only ones to have crossed the NOMA threshold. The emergent field of social cognitive neuroscience uses new brain imaging methods like fMRI to study how the brain responds to social interactions. This relatively new field has unified previously separate studies of the function and structure of the brain, typically the domain of neuroscience, and theories of mind or belief, typically the domain of philosophy, psychiatry, and religion.

The fall of NOMA may in part reflect the changing demographics of US scientists. Despite a historically vocal group of scientists like Dawkins and Lewontin who view religion as incompatible with science, 39 percent of scientists in the US today identify with a religious tradition, and 70 percent of US scientists see no conflict between science and religion. Yet even today, long after the evolution versus creationism debate has subsided in America, within many labs a separation of church and state remains. Even as the study of belief has become more accepted, scientists with religious affiliations still compartmentalize their spiritual and professional lives and rarely refer to their beliefs as motivating their experiments.

But with NOMA no longer standing unchallenged, new research questions become possible. Why study only the effects of Buddhist beliefs and meditation practices on the brain? Why not study the biological effects of Christian beliefs and prayers? Or the impact on the brain of Muslim or Jewish practices? Surely, given what we know about the effect on the brain of gratitude and meditation more generally, the influence of other religious and spiritual beliefs and related practices is also worth examining.

There's a distinction between scientific studies of the effects of spiritual beliefs on well-being and studies aimed at proving the truth of God's existence. The latter is a question for theologians. Studying spiritual beliefs need not entail "abusing science" to support religion. The question of whether there are empirical effects at work is simply that: an empirical question that can be tested.

An exciting new field of study now lies within our reach on the neurobiological effects of spiritual beliefs on our brains, our bodies, and our health. There is so much more to explore. Belief, as we'll discover next, is a key building block of our mind and brain. Let's turn to what we now know about how beliefs form the basis for perception and cognition—and the fundamental inseparability of our beliefs from our biology.

In the next two chapters, our focus will be on beliefs generally, not just on spiritual beliefs per se. The scientists who laid much of the neuroscientific groundwork for beliefs used a more secular lens, and I draw on that literature here. The overall goal of the next two chapters is to show how and why our beliefs are inextricably linked to our biology. I will return to more explicit connections between spiritual beliefs and health in part 2.

2

How Beliefs Affect the Brain and the Brain Affects Health

My paternal grandfather, after serving in World War II, became a metalworker at Lockheed. He set up a full shop in a detached garage on his and my grandmother's farm in Northern California. I remember spending summers there, watching and trying to emulate him by working with the table saws, drill presses, and lathes. I have so many positive memories from that time that readily come to mind: the satisfaction of completing a project and the smell of building things—freshly sawn lumber, latex paint, and the smoky aroma of solder used for plumbing and electronics projects. My memories confirm my belief that my grandfather, like me, loved tools and building things—and that he was talented at his craft.

A belief is a conviction about what we hold to be true. We form beliefs in part through introspection or memories. What I remember about my grandfather informs my beliefs about him. Belief can be also based on sensory input. One whiff of cinnamon and you'll probably believe it is spicy and delicious. Finally, our beliefs are informed by communication with trusted others. Those who believe that we should treat others as we would like to be treated typically learned this from others, as I did, at an early age from parents, teachers, or religious leaders. These beliefs collectively form the foundation of how we interpret and interact with the world.

To understand scientifically how beliefs affect our health, it's helpful to break the process into two parts. First, how do beliefs affect the brain? Second, how does the brain affect our overall health? We consider these below in turn.

Individual beliefs exist within a larger framework of beliefs, or a *cognitive map*. This term was coined by psychologist Edward C. Tolman in the 1940s. Tolman, the son of a successful businessman and a homemaker interested in education, initially studied electrochemistry at MIT. While a student there, Tolman began reading the work of William James, a professor of psychology at Harvard University in the late nineteenth century. James is generally credited as the "father of American psychology." After reading James, Tolman discovered his true passion. He switched fields and did his graduate studies at Harvard in psychology. Tolman used mazes to study learning and behavior in rats. During his experiments, Tolman realized that the rats were able to form a mental representation of the mazes he placed them in. When food was consistently placed at the left end of a Y-maze and water at the end of the right arm, hungry rats dropped in the maze statistically turned left more often, and thirsty rats statistically turned right more often. Through previous runs, the rats had acquired a cognitive map that food was to the left, water to the right.

Our cognitive map is more than just a mental representation of a specific area. It's a subjective model of how the world works. If we drive down the road and see a red light, we stop. Our belief that red means stop is part of a larger framework for how red lights, green lights, and traffic operate. This framework isn't just a set of objective or societal rules, however. It's a very personal mental blueprint. Imagine a local supermarket. That store has a specific address that anyone can use to locate it. In our cognitive map, however, that store has associations that are unique to us. It might be the place where we blissfully encountered our first comic book. Or maybe we became a regular customer after discovering it carried a favorite candy. Later, maybe it became somewhere to avoid after an incident where we felt embarrassed in front of someone we liked. Cognitive maps encompass spatial relationships, sensory impressions, and personal associations. Our maps represent the entirety of our social and physical world, offer an explanatory structure for that world, and guide us in how to interact with the world.

When we perceive something in the world at any particular moment, we try to fit that evidence into our cognitive map to generate a prediction about what will happen next. When I was first diagnosed with a brain tumor, I predicted that while my exact survival time might vary, I had as little as a few years left to live. My cognitive map, or my larger model of how the world works, generated additional beliefs and predictions based on that timeline. For example, I predicted that my premature death would leave my wife, Candy, as a single mother with a demanding job, in difficult straits, and that some of my scientific publications might not see the light of day. These predictions in turn guided a set of actions: to ensure that my will was in order and my life insurance up-to-date so that Candy and our daughter would be taken care of, and to complete my lab work as quickly as possible.

Cosmologically, I also made a set of predictions about the future based on the preexisting beliefs that made up my cognitive map. I believe, as Genesis states, that "God breathed life into Adam" and "he became a living soul." I interpret this to mean that we are souls, and because a part of myself is eternal or soul, it can survive death. That in turn led me to a new set of predictions: that death is not final, that death is not something to fear, at least ultimately. These predictions affected my brain—and thus my emotions, helping me maintain a more positive outlook.

My diagnosis triggered these cognitive processes, which in turn led to a set of predictions about what would happen next. These cognitive processes are not just thoughts: They are biological processes in my brain. We can thus trace a basic trail. Our larger framework of beliefs or cognitive map affects our cognition or perception, which in turn affects our emotions, which in turn affect our health.

Let's start with how our beliefs affect our cognition or perception. The basic task of the brain is to make sense of the reality that it encounters so that it knows what action it needs to take to survive. It all starts with a prediction.

Predictions and Reality

Hermann von Helmholtz, a nineteenth-century German physician and physicist, was one of the first to point out that we perceive reality by making predictions, not by reference to facts. When we perceive an object, according to von Helmholtz, the brain makes a series of predictions on the probable causes of the sensory input we've just received.

Suppose your brain receives sensory input of a round, sweet-smelling orange object on a distant table and makes a prediction. It attempts to match its sensory input—round, sweet-smelling, orange-colored—to its prediction of what the object truly is: likely a piece of fruit, probably an orange.

But capturing reality is easier said than done. The brain is often in a state of uncertainty. Take the case of our round orange object. Perhaps, as we approach the table and touch the fruit, the object's outer shell turns out to be soft, not hard like an orange rind. Also, it smells sweet and fragrant in a non-orange way. These new sensory inputs have added information and registered an error in the initial prediction. A new prediction is required. This is not an orange. Even better—it's a summer peach.

The brain will continue automatically to make predictions about the reality of the fruit on the table until it stops receiving significant new or additional information. As von Helmholtz observed, the brain is a prediction or inference machine. Like a chef who continuously adjusts the amount of salt in a recipe, tasting and further adjusting the quantity of the seasoning to achieve the dish's best flavor, the brain continuously generates predictions until it lands on a prediction that matches reality. The brain doesn't stop trying to make more predictions until the chance of error is small enough that it's basically trivial. A 100 percent match of prediction or perception to reality may or may not be reached, but that's the goal. At some point, our brain's certainty that it has correctly identified and modeled the reality of what's before us (say 90 percent) will be great enough.

Predicting whether the fruit is an orange or a peach does depend on some previous knowledge of both oranges and peaches. Let's say we're walking down the road on a foggy night in January, and we see a shape in the distance. Is it a coyote or is it a dog? Our prediction has real consequences for what we decide to do, and so our brain will try to be as accurate as possible, automatically working to assemble additional sensory input. If the animal's posture is hunched, not straight, we may predict that it is slightly more likely to be a coyote than a dog. Is it barking or howling? What size are the ears?

Beyond the sensory input, however, we will also rely on past experiences or beliefs about how likely it is to encounter dogs or coyotes on this route. Beliefs are themselves predictions. To clarify, they are not a prediction about a specific event such as the encounter we are now having in January with the coyote/dog. Rather they are predictions about how likely, in general, something is to happen. A belief is a probability, not a certainty. What we believe might not be and/or might not happen. Yet we act on and respond to those probabilities as if they were certainties. And beliefs inform predictions in specific events such as our (hypothetical) January encounter.

Belief about how likely it is to encounter either a coyote or a dog on the route gives one of those animals a head start in the competition—dog? coyote?—in our brains. If we more often see coyotes on this road, the coyote prediction has a head start, and if we usually see dogs, the dog prediction noses ahead.

The Neuroscience of Beliefs

Helmholtz's nineteenth-century theory of the brain as a prediction machine was just that—a theory. Thanks to modern neuroimaging methods and studies in neurophysiology, however, we can now pinpoint how the prediction process works biologically.

We've noted that the brain is made up of roughly 100 billion neurons, which are cells that transfer information around the brain. Neurons each make an average of a thousand connections with other neurons, which is the basis of signal sharing. Each neuron has a voltage inside, like a battery. When a neuron receives an activation signal, such as the chemicals in the smell of an orange, the voltage inside the neuron rises or spikes.

In our orange example, the primary neuron activated is a sensory neuron. That sensory neuron activates other neurons, such as those in the olfactory bulb that register the orange smell, or those in the *orbital frontal cortex* (a region in the frontal lobe of the brain) that signal both what the object is and whether it is valuable—or worthless.

Scientists know which neurons in which parts of the brain are being activated because we can measure the different activity levels or voltage spikes of individual cells with tools such as electrodes. The more active the neuron is, the more spikes it generates per second. Every sensation we perceive, every prediction our brains make, and every action we take are either a result of, or signaled by, voltage spikes in neurons in the brain that can be measured. Years ago, when I was studying the neurophysiology of animal brains, I found I could predict what action the animal was going to make, before the animal made it, *just by measuring the activity of its brain cells.*

Activity in a cell and/or patterns of activity across multiple cells signal that a particular event is happening. Going back to our coyote or dog example, if we see an animal in the distance and it looks more like a coyote, then the set of "coyote neurons," or the cells that predict that what you're seeing is a coyote, are the most active. It will seem most likely to us that we are looking at a coyote. Conversely, if the activity pattern is stronger in the set of "dog neurons," it will seem more likely that we are looking at a dog.

What's important to understand is that every prediction has a biological basis. The decision whether the animal in the distance is a coyote or a dog is not just an abstract puzzle. *Every prediction is both based in—and potentially changing—not only our perception of reality but also our biology.*

Emotion Neurons

Just as there are specific dog neurons and coyote neurons that get activated when we see a dog or coyote, those neurons in turn trigger specialized "emotion neurons." Which emotion neurons activate depends on what we're perceiving, the context of the event, and our beliefs about the situation or cognitive map. If, for example, we believe that dogs are generally friendly and we like dogs, seeing a panting St. Bernard approach will activate what I'll call positive emotion neurons. These neurons, associated with feelings of joy, happiness, and pleasure, are mainly located in the *ventral medial prefrontal cortex* (the lower middle front part of our brain) and the nearby *ventral striatum*.

Conversely, if our experience with dogs in the past was traumatic, our cognitive map may lead us to predict that the panting St. Bernard is a threat. In that case, our dog neurons will activate neurons that signal danger. These inverse "negative emotion neurons" appear in different brain regions. Negative emotion neurons such as those associated with fear, rejection, and disgust are located in the *insula*, or the region between the frontal and temporal lobe, and in the amygdala, the part of our brain that triggers such physiological impacts of fear as a racing heart. We may also activate neurons that signal how much we like or don't like something, which are located in the *medial prefrontal cortex* or the area in the mid-front of the brain. The neural response to fear is not determined only by experiences. Genetic differences among people can also influence the intensity and duration of each person's emotional responses. For example, a genetic variation related to the serotonin receptor can cause increased amygdala activation for fear, and another genetic variation that controls dopamine processing can make it harder to unlearn fear, even when we know it is safe.

Downstream Effects

Emotion neurons have physiological downstream effects—including, ultimately, on our health. Specifically, they trigger a communication system called the *hypothalamic-pituitary-adrenal (HPA) axis*. The hypothalamus and pituitary are two small organs next to each other at the front base of the brain. They are like the president and vice president of the HPA axis. Together with the adrenal glands, or the middle management, which sit on top of the kidneys, the hypothalamus and pituitary regulate body processes such as digestion, temperature, mood, sexual activity, and immune response.

The hypothalamus acts as the control center with a goal of maintaining homeostasis or a static state. Think of a thermostat. Because the thermostat's goal is to maintain the desired temperature or set point, when it receives information that a room's temperature has exceeded that set point, the thermostat triggers the cooling system, and when the room temperature drops below the set point, it triggers the heating system. Likewise, temperature-sensitive neurons in the hypothalamus control body temperature. When body temperature drops, among other things, the hypothalamus signals the sympathetic nervous system to release norepinephrine, a hormone that causes constriction of blood vessels, which reduces heat loss through the skin. In addition, the hypothalamus may trigger the release of thyroid hormones that increase metabolic rate and subsequent heat production. If the body overheats, the hypothalamus facilitates cooling by triggering evaporation or sweating and the dilation of blood vessels, which increases heat loss through the skin.

The hypothalamus similarly maintains a static state when it comes to hunger, triggering a drive to eat when blood sugar drops or feelings of satiety when we have ingested sufficient calories to restore energy balance. Similarly, the hypothalamus regulates other basic needs, like thirst, sleep, and sexual activity. The pituitary gland participates in the process of homeostasis by releasing hormones that maintain desired states, including those that drive testosterone production and thyroid hormones, which increase or decrease metabolism rate.

What could be more important to equilibrium or steady state than staying out of harm's way? The HPA axis plays a key role here as well. Imagine we are being chased by a hungry bear. When we perceive danger, we activate our "fear neurons." From there, a long chain of communication ensues. Most immediately, the hypothalamus, through the HPA axis, triggers the sympathetic nervous system to release adrenaline from the adrenal glands. Adrenaline increases the heart rate, opens the airways in the lungs, dulls pain, and increases blood flow to the muscles so that we can take flight—or fight back.

Adrenaline isn't the only weapon in the HPA axis's arsenal, however. Once the amygdala signals the hypothalamus that danger's afoot, the hypothalamus seeks to increase our *cortisol* levels. Again, this involves a long chain of command. The hypothalamus, as the president of homeostasis, signals its vice president, the pituitary, via corticotropin-releasing hormone (CRH), to release a different hormone, adrenocorticotropic hormone (ACTH), in the blood, which triggers the adrenal glands or middle management to release cortisol. In turn, cortisol increases sugar or glucose in the bloodstream and the brain's use of glucose so that we have more energy for running away from the bear. It also suppresses many of the other functions in our body that use energy, like digestion and immune function, so that all our energy can go into self-defense.

The hypothalamus has its own cortisol monitors (glucocorticoid receptors), and if they detect high levels of cortisol circulating in the body, they will, through the HPA axis, slow the release of additional cortisol, to maintain a steady state. Like Goldilocks and her porridge, the hypothalamus prefers homeostatic states, and thus cortisol levels, to be just right. If only our emotions—and stress levels—would cooperate.

Health Implications

You've probably already heard about "good stress" and "bad stress." Bodies are made for a certain amount of stress—and cortisol. We want to be able to produce adrenaline and cortisol if we're facing danger—or even a difficult math test. But overproduction of cortisol can have negative effects. Especially if our body is consistently stressed.

Chronic stress is associated with a range of health conditions, from depression, cardiovascular disease, diabetes, and upper respiratory infections to autoimmune diseases. The details of stress's effects vary by disease, but it's safe to say that in many cases, something has gone wrong with the HPA axis's function, and hormones may be flooding the body as a result. Since excess cortisol is one of the main consequences of chronic stress, let's look at how that affects the HPA axis—and thus our health.

We've seen that when we experience a stressor, the hypothalamus produces CRH or the hormone that triggers the pituitary gland to release ACTH, which triggers the adrenal glands to produce cortisol. The hypothalamus, like a thermostat aiming to keep the temperature at set point, tries to keep the levels of cortisol relatively constant (though there are normal fluctuations throughout the day). But if too much cortisol gets produced in the body, the hypothalamus loses its sensitivity to cortisol and baseline cortisol levels rise. It's as if the thermostat broke.

But the body also keeps its checks and balances: Another thermostat exists in the room, that also registers temperature. The secondary thermostat, which accurately senses that the room has become too hot, can tell the less reliable main thermostat to reduce the flow of heat. In the case of the brain, this double-check function is fulfilled by the *hippocampus*, a brain region involved in memory. The hippocampus can send an inhibitory signal to the broken hypothalamus, telling it to reduce the flow of cortisol to the body. Unfortunately, high levels of cortisol circulating in the body over time, due to chronic stress, can also damage the hippocampus, shrinking it. The shrinkage impairs the hippocampus's ability to signal the hypothalamus to send its negative feedback. It's as if the backup thermostat is also broken. Instead of signaling the main thermostat to turn off the heat, the secondary thermostat does nothing. As a result, the furnace keeps blasting hot air, or cortisol, into the system.

Excess cortisol over a longer time has a number of adverse health effects. Cortisol increases blood glucose, which is extra fuel for muscles in the short term, but over a longer term can contribute to diabetes and weight gain. Cortisol also makes blood vessels more sensitive to signals that constrict the blood vessels and causes the heart to beat more strongly, which can cause high blood pressure and increase risk of heart attacks and strokes. Too much cortisol reduces bone growth and maintenance, which over time can cause osteoporosis. Similarly, cortisol redirects energy away from the immune system, which over time makes us more susceptible to severe infection, as it weakens the parts of the immune system that fight bacteria, viruses, fungi, parasites, and cancers.

Negative emotion neurons have other downstream physiological effects, though much of what we definitively know about health effects circles back to stress, the HPA axis, and cortisol. On the neural pathways side, we could look beyond the hypothalamus and broader HPA axis to the sympathetic nervous system. When overactivated, the sympathetic nervous system may increase production of inflammatory cells. This excess inflammation damages other cells throughout the body and can cause leptin resistance, which results in excess weight gain. On the hormone side, other hormones do get involved, but the link is often through cortisol. Stress elevates cortisol levels, which in turn elevate hormones such as ghrelin and reduce the brain's sensitivity to leptin. The end effect is increased appetite and decreased satiety. As these hormones become dysregulated, it affects how much we eat and thus our weight. Similarly, elevated cortisol levels affect aldosterone, a hormone that affects blood pressure. Dysregulated aldosterone can lead to high blood pressure, kidney disease, and heart failure.

The Full Circle

The gap between possessing a set of beliefs and exacerbating one's heart disease or even cancer might have sounded laughably large at first. Hopefully, however, it's clearer now that this gap isn't as wide as originally thought. Our cognitive map or set of beliefs leads us to make predictions. Those predictions trigger specific physiological activity, and emotions, in the brain. The resulting physiological processes triggered by those thoughts and emotions—positive and negative—can then affect our health.

Relationships among belief, experience, and physiological response are not one-way. If an upsetting event triggers a disruption in our homeostatic or steady state, affecting our hormone levels and health, it can also change the connections in our brain regarding the strength of that event's memory, which in turn can affect our cognitive map. This is especially true with traumatic experiences.

The attacks against the United States on the morning of September 11, 2001, deeply affected me. Such attacks on US soil were once unthinkable in my lifetime. In the aftermath, my inherent sense of security and safety was shaken. We now know that terrorist attacks are very rare in the US, but in the days after 9/11, that was unclear. In the back of my mind, I began to worry that a similar attack could injure or kill me or my family.

What happened to me also happened to many others—children and adults—on and in the wake of 9/11. It's what we call a *flashbulb memory.* Flashbulb memories are exactly what they sound like—we react as if a flashbulb has gone off near our face. The memory is detailed and vivid, often because of the intense surprise and shock of the event. We retain that event much longer than others as we are literally rewiring connections around it in our brain, bolstering that memory over others. And retaining that memory—and the negative emotions associated with it—affects our prediction of future events.

One week after the 9/11 attacks, news reports generated another scare, this time involving anthrax spores deliberately mailed through the post office to public figures. The spores appeared as a fine white powder that spread as easily as powdered sugar but were full of deadly bacteria. Inhaling even a tiny amount could cause death as the bacteria grew quickly and filled the lungs. Worse, the anthrax-filled envelopes could leak while in transit through the postal system, contaminating other nearby mail. Potentially any letter could have anthrax, it seemed. I knew intellectually that the odds of *my* mail carrying anthrax were almost zero. But with vivid memories of 9/11 still lingering in my mind, my heart rate and breathing accelerated every time I checked the mailbox. My brain was predicting that a terrorist act, this time involving anthrax, might hurt or kill me.

Months went by, the anthrax assaults stopped, and it eventually became clear that another terrorist attack on the same scale was unlikely. This should have changed my predictions about world events, but the psychological and physiological effects lingered for many months after. The traumatic experience of watching the World Trade Center Twin Towers fall had formed a flashbulb memory. I retained negative perceptions of my safety in my cognitive map even after the conditions had changed. And my heart rate and breathing still quickened at the mailbox.

Remember that we said our beliefs, and thus our cognitive maps, are formed by sensory impressions, memories, introspection, and communication with trusted others. My sensory impressions of the attacks, my flashbulb memory, and the larger model of the world that the attacks conveyed all came together with elements of my genetic makeup and experiences, to form my model of how the world works. We'll come back to how and why stress can affect mental health as much as physical health in part 2, but the fundamentals are not dissimilar to the HPA axis and cortisol chain of events outlined above.

The next chapter explains how we form and update our beliefs, including cognitive biases that can lead us astray—as well as ways to work around them. Understanding how we form beliefs can in turn help us find what is both true and good for our health.

3

How We Form Beliefs

THE JURY WAS SEATED FOR THE CRIMINAL TRIAL. THE CHARGE was first-degree murder. Frank Johnson, the defendant, was accused of killing Alan Caldwell. The two had argued earlier in the day before Caldwell's death. That evening outside a bar, they got into a fight; the evidence was clear that Johnson stabbed Caldwell, but it was in dispute whether Johnson lunged at Caldwell with the knife or whether Johnson only held out the knife to protect himself from Caldwell's attack. It was also unclear whether Johnson went home to get the knife and then came back, and whether he went to the bar to find Caldwell. The jurors listened as the prosecution and defense took turns arguing that Johnson had planned to kill Caldwell after their earlier argument, or that Johnson was afraid of Caldwell (due to Caldwell's large, intimidating size), and kept a knife only to defend himself. The witnesses gave conflicting testimony. Finally, the judge gave the jury solemn instructions, and the jurors began deliberating.

"Johnson went to the bar with his knife, hoping to find Caldwell to kill him. When they were both outside the bar,

Johnson took his opportunity and killed Caldwell," said one juror.

"Johnson was angry, but he didn't want to kill Caldwell. He carried a knife for self-defense, and when Caldwell moved toward him to attack, Johnson held out the knife, accidentally stabbing Caldwell in self-defense," said another juror.

The two jurors are not simply arguing with each other. Rather, they are telling different *stories*. One is a story in which Johnson is guilty, and the other is a tragedy, but not first-degree murder. If a majority of jurors believe the first story, they will likely convict, but if they believe the second story, they will acquit. How do the jurors know which story to believe?

In the 1980s, Nancy Pennington and Reid Hastie, both professors of psychology at the University of Colorado Boulder, studied how jurors form beliefs using the fictitious case of *Massachusetts v. Johnson* described above. They found that when jurors hear evidence, they try to fit the pieces of evidence together into a story. First this happened, and then that happened, and then the killing happened, and so on. Different jurors may construct different stories from the same evidence. Each juror may entertain multiple plausible stories. Ultimately though, they have to believe one story as more plausible in order to reach a verdict. How do they decide? What Pennington and Hastie found is that jurors form beliefs about which story is correct based on two criteria: coverage, or how well a story accounts for all the facts; and coherence, or how well a story fits together with itself and preexisting beliefs. If a story casts Johnson as killing in self-defense but can't explain why the surveillance video shows Johnson lunging at Caldwell, then the story lacks coverage of the evidence. Similarly, if the evidence suggests that Johnson was preparing to take his kids to a ball game the next day but also planned in advance to kill Caldwell, the story lacks coherence or internal consistency. Jurors believe the story they can construct that has the best coverage and coherence, and they believe the story more strongly if it fits the evidence uniquely better than competing stories.

Making it easier for a juror to construct a story that favors one verdict over another has a disturbingly large effect on final verdicts. Attorneys at trial can choose in what order to call witnesses, deciding whether to tell a story of events in chronological order, or instead to call witnesses to address specific legal questions. When the prosecution presents its story of events in chronological order while the defense presents evidence in nonchronological order, jurors convict 78 percent of the time. When the converse is true, and the defense presents its story in chronological order but the prosecution presents in nonchronological order, the jurors convict only 31 percent of the time. Even when the facts of the case are the same! Presenting those facts in a way that makes it easier to form a coherent story can determine whether trial outcomes favor the defense or prosecution.

This is how we form beliefs about everything from whether someone committed a crime to whether spiritual practices can heal us to where the universe came from, by constructing stories about what has happened in the world. These beliefs collectively form our cognitive map. The process begins with evidence. Sometimes, that's from direct experience, like the smell and taste of a peach. Other times, the evidence comes from people who convince us that something is true. In the courtroom, jurors hear from witnesses and attorneys who present evidence and arguments. The jurors, and anyone else listening, will be persuaded to believe an account if four factors of persuasion are present: First, is the witness credible, such as an expert or eyewitness who seems reliable? Second, is the testimony plausible? Did the witness claim something believable such as that Johnson lunged at Caldwell, or did the witness claim that Johnson did a double backflip before stabbing Caldwell? The former would seem reasonable; the latter not so much. Third, is the witness talking in a credible context? If under oath in a courtroom, and under penalty of perjury, that would seem more credible than an informal conversation in a back room. Fourth, is the audience amenable to the message? If, for instance, the jurors happened to be trained lawyers themselves, they might be more likely to spot inconsistencies in a story. If these four factors are present, the audience is more likely to be persuaded.

Our spiritual beliefs form similarly. We are presented with sensory experiences and evidence from those we trust, and we try to piece together a cognitive map of the world that covers the evidence we have, is internally consistent, and (ideally) seems uniquely better than alternative stories. Like courts, we judge the evidence and construct the best story we can. And like jurors rendering verdicts, we act in ways that seem to provide the best continuation of the story, living by moral or ethical codes and participating in spiritual practices that fit our beliefs.

For better or worse, our brains are wired to form and maintain spiritual beliefs. One leading neuroscience researcher in this area, Jordan Grafman, is a friend of mine, an amiable man with an infectious smile, and an incredibly prolific scholar with hundreds of scientific publications to his name. A former section head in Cognitive Neuroscience at the National Institutes of Health in Bethesda, Maryland, he studied the frontal lobe and a number of clinical disorders that affect it. Later, as the nonoverlapping magisteria (NOMA) sensibilities fell out of favor in the mid-2000s, Dimitrios Kapogiannis joined his lab. Kapogiannis wanted to study how religious beliefs work in our brains. This sparked Grafman's interest. He had previously been curious about how the brain processes political and economic beliefs, and was amazed that so little work had been done to explore religious beliefs in the brain. He became, as he told me, "infatuated" with the idea of studying how our brains process religious belief. He has been doing so ever since.

In their study, Kapogiannis and Grafman first formed a set of yes or no questions. Participants in their experiment, some of whom were religious and/or spiritual and some of whom were not, had to indicate whether they agreed with a series of statements regarding beliefs about God. The questions fell into three categories: namely, God's perceived level of involvement (e.g., "God is removed from the world"), God's perceived emotion (e.g., "God is angered by human sin"), and how much people based their views on direct spiritual experiences versus more abstract religious doctrines (e.g., "God dictates celebrating the Sabbath").

Grafman and Kapogiannis put a subset of participants in the fMRI scanner and asked them to rate their agreement or disagreement with statements about God. The results were surprising. When the experimenters asked participants questions about God's perceived lack of involvement, several brain regions became more active. These included regions in the prefrontal cortex, both left and right, which are involved in higher cognitive function, and regions of the right temporal lobe and occipital lobe, which are involved in recognizing objects and visual imagery. Counterintuitively, questions about God's active involvement in the world did not lead to increased brain activation. Statements about God's love activated the prefrontal cortex just behind the forehead, while statements about God's anger activated part of the temporal lobe. Statements about religious doctrines activated the temporal lobe, while statements about direct spiritual experiences strongly activated the visual cortex, which processes what we see as well as visual memories and imagination.

What these results show is that many different parts of our brains are wired for belief, including beliefs about God or other unseen spirits or energies. We may believe in God or another higher power, or we may believe there is no God or that God is uninvolved in our lives, but either way, *we have spiritual beliefs, and our brains carry concepts of God*.

To be clear, I'm not saying that Kapogiannis and Grafman's results prove that God exists in our brains, or that God is directly manipulating our brains. As a Christian, I personally believe that we may at times experience God directly, but these data don't prove that—they show only that our brains have the capacity to process beliefs about God. Think about it this way: Suppose our loved one leaves us a special gift, and we find it. We see that our loved one has done something special for us. We feel happy. Inside our brain, our visual cortex processes what we see of the gift, and our temporal lobes have neurons that fire when we see, or think about, our loved one. They start firing, along with other neurons that fire to signal that our loved one did something nice for us. These cells, as part of our cognitive map, collectively signal our belief that a loved one indeed loves us. That thought, and our belief about our loved one, literally consists of the neural activity pattern. Put another way, if some of those brain areas were removed, we would be incapable of having those thoughts. People with damage to parts of the frontal lobe may retain their intelligence, but they feel no empathy for people they hurt and are completely incapable of feeling regret (these people usually end up behind bars). In the same way that neurons in our temporal lobe get activated when we think about our loved one, there are other neurons that get activated when we think about God. They are part of our cognitive map and allow us to think, reason, and hold beliefs about God.

Some researchers argue that religious beliefs are a by-product of evolution. In this view, our ability to form beliefs about God is an outgrowth of our ability to form beliefs about other people and even involves the same brain regions. Pascal Boyer is a professor of psychology at Washington University in St. Louis, in the same department where I was a postdoctoral fellow. He argues that religious beliefs and rituals grow out of our need to keep order and safety in our lives, which leads to religious practices aimed at purifying and protecting, such as washing hands before prayer or burning sage. Similarly, Richard Sosis, a professor of anthropology at the University of Connecticut, and his doctoral trainee Candace Alcorta note that shared beliefs can help unite people to a common cause, even when they don't share bonds of kinship. Religious communes survive longer than secular communes, at least in part because a shared spiritual framework improves social cooperation. We are predisposed to form religious beliefs, Sosis and Alcorta argue, because our religious ancestors cooperated and survived better than their nonreligious peers.

Religious beliefs may help us to cooperate, but why we hold our beliefs has as much to do with emotion as with reason. Remember that we base our beliefs not only on ideas but also on how experiences make us feel. Brain imaging studies suggest that we believe what we like and disbelieve what we don't. In the study by Grafman and Kapogiannis described earlier, when people disagreed with religious statements, there was more activation of the insula and cingulate, two areas associated with unpleasant emotions and a desire to avoid them. Another brain imaging study, by noted leader of the New Atheism movement Sam Harris during his PhD studies at UCLA, found the same result: When people evaluate statements they disbelieve, the cingulate and insula are activated, but when people evaluate statements they believe, the ventromedial prefrontal cortex is active instead. This area is primarily associated with pleasure and is more active when people get something they like more.

When it comes to beliefs, each of us is our own jury deliberating to reach a verdict about what is true. Religious and/or spiritual beliefs are true, in our judgment, to the extent that they explain the world, and to some extent whether we like that explanation. Our brains are constantly working to find beliefs that make sense of the world and our place in it. Professor William James argued that the value of our beliefs is not so much that they are objectively true but rather that they provide a sense of meaning and purpose that enriches our lives. In the end, we like and value our beliefs because they enrich us. Still, people can and do change their beliefs.

How Beliefs Change

It was a pleasant, sunny morning in the middle of March 1992. A sophomore at the University of California, San Diego at the time, I was deeply engrossed in mechanical engineering homework and just about to leave my dorm room for my next class. UC San Diego ran on a quarter system, not a semester system, which meant that final exams for the winter quarter were around the corner. In just a week, I could go home for spring break and enjoy a welcome respite until the next term began. I was looking forward to seeing my parents and sister for the first time since Christmas.

And then my phone rang.

"Josh, it's Mom. Can we have lunch today?"

I was surprised, as my mother didn't typically initiate a visit with me at the university, especially on short notice. A few hours later, we sat down to lunch in the Revelle College dining hall. Then the bomb dropped.

"Your father and I are getting divorced," she said with a look of weariness that conveyed how much she had cried in the last few months since I last saw her.

"We decided shortly after Christmas, but we waited to tell you until you finished your classes, so as not to distract you. I moved out in January."

I sat speechless for a moment, looking across the dining hall table at my mother. My world, as I knew it, had changed, and it was going to take all of us in the family a long time to process.

What I experienced in that moment was surprise and distress. Sometimes surprises can be pleasant, such as when my grandparents gifted me a car when I was in high school. But pleasant or not, surprises drive changes in belief. They tell us that our predictions were wrong because our beliefs are wrong. Our cognitive map needs updating. So, we search for a new and better story to make sense of the world, and we update our beliefs accordingly. Or rather, our brain updates our beliefs. In my case, my beliefs about the stability of my family were upended and replaced with a new belief that families are fragile. Although I knew that both my parents loved me, I now fully believed that my family interactions would look very different going forward.

Surprise is a powerful signal in the brain. In 2009, a postdoctoral fellow in my lab, Derek Nee, and I were looking at some fMRI data we had just collected to determine how brains switch tasks. Derek, now a professor at Florida State University, has a razor-sharp wit and is excellent at spotting nuances in data that would otherwise easily escape notice. Derek noticed something, well, surprising. Whenever our participants experienced something surprising (usually because they made a mistake), the anterior cingulate cortex and insula became active. Some events were more surprising than others, because they occurred less often. The most surprising events generated the strongest activation. In later work, we showed that the surprise signal seemed to happen because the brain processed, among other things, two kinds of signals—one signaled what it expected to happen, and the other signaled what actually happened. When those patterns were different, the cingulate and insula activated to indicate a conflict between predicted and actual outcomes.

Surprise, it turns out, drives the brain to update its beliefs. In 2015, a research group was studying patients with brain damage from strokes. Two patients had damage to the same brain areas, except that one also had damage to the anterior cingulate, and the other did not. The only behavioral difference between them was that the patient with anterior cingulate damage could not learn to do a task when the rules changed. The area that signaled surprise was broken, so other parts of the brain that needed that surprise signal to drive learning could no longer adapt.

Surprise causes our brains to update beliefs, but sometimes we only later become aware that we've done so. In 1997, Professor of Psychology and Neuroscience Antoine Bechara and colleagues at the University of Iowa were studying how people learn to perform a task that became known as the Iowa Gambling Task. The task was simple—choose a playing card from one of four possible decks. Each card has a reward or punishment amount on it. Participants win the amount of reward on the card or, if the card carries a punishment, lose the amount of money on the card. What the participants weren't told was that two of the decks were better than the others. The "bad" decks initially seemed more attractive—most of the time, they paid a larger reward, but occasionally they inflicted an even larger loss, so that over time the bad decks caused subjects to lose money. The two "good" decks initially seemed less attractive, because they usually paid a smaller reward. Still, they were better because they gave very small punishments, so that over time the participants gained money by choosing from the good decks. Subjects initially noticed the larger reward in the bad decks and picked up lots of bad deck cards, until they started getting some big losses. Then they switched and started picking up more cards from the good decks.

To understand the game better, imagine standing at a high-value slot machine in Las Vegas, with lots of blinking lights and flashing colors drawing us in. We are about to pay one hundred dollars to pull the lever. We feel a bit of excitement and apprehension, and perhaps some tingling as we flush with the realization that we could win big—or lose one hundred dollars. That tingling happens as our brain makes predictions, and emotion neurons start to activate our fight-or-flight response. In addition to adrenaline and cortisol release, our skin starts to sweat a little as our breathing and heart rate increase. That sweat actually decreases the skin's electrical resistance. In the Iowa Gambling Task, Bechara identified this electrical resistance as a way of measuring emotion, or more specifically, apprehension about losing money. While participants played the card task, Bechara found that after a few losses from the bad decks, they started to show a strong electrical resistance drop just before picking a card from the bad decks. Their brains had learned that the bad decks were in fact bad and would cause them to lose money, and they started picking more cards from the good decks.

Meanwhile, every few card draws, the experimenters asked participants, "Tell me all you know about what is going on in this game." The participants said they knew nothing. All the decks seemed the same to them. Still, every time they were about to choose from one of the bad decks, their skin resistance decreased, even if they didn't lose money on that card draw. Their brains were learning the nature of the task and which choices were worse, even causing them to have an emotional reaction and sweat a little. This happened even though the subjects were not aware that they were learning!

Have you ever met someone and immediately had a bad feeling about them, without knowing why? Maybe you sensed that you should avoid them, but you also felt badly because there was nothing obviously wrong with them. In those situations, our brain has learned to pick up on subtle cues that signal something might be wrong, even if we are not consciously aware of why we have a bad feeling. Our brain is looking out for us.

As it turns out, a very specific part of our brains is learning to look out for us. The Iowa Gambling Task got me thinking about where in the brain that might be. In 2005, I proposed to my then-postdoctoral adviser, Todd Braver, a follow-up experiment. We used brain imaging to identify which parts of the brain were learning without the participants being aware of it. Todd is a brilliant, highly productive scientist with a laid-back attitude that fit well with his former days as a surfer in San Diego. Todd gave me a fair amount of freedom to create my own projects, so I designed an experiment with him to challenge a popular theory of how the anterior cingulate cortex works. In the end, I found evidence consistent with a newer theory I had proposed, but something else in the data also caught our attention. We found that when participants were about to make a mistake, their anterior cingulate became more active, even though at the time, subjects told us they had no idea when they were more likely to make a mistake. Our anterior cingulate is learning from surprise, even when we are not aware of it.

The learning our brains do is the bedrock of *neuroplasticity*. Much has been written about neuroplasticity and its benefits, but apart from the mystique, it is simply a change in how strongly neurons connect with each other. If like Pavlov's dogs we hear a bell ring, and then some food appears, the "bell" neurons in our brain become active, and then the "food" neurons become active. When they are active together, the connection from the bell neurons to the food neurons gets stronger (through a process called long-term potentiation, or LTP). Later, when we hear the bell ring, the bell neurons activate the food neurons more strongly, and so we lick our lips because dinner is about to be served. In 2000, Dr. Eric Kandel, a researcher at the Howard Hughes Medical Institute, won the Nobel Prize in Physiology or Medicine for helping us understand how this process works, because it is at the core of how we learn.

There is more to neuroplasticity, though, and this is where surprise comes in. When we can perfectly predict what will happen next, there is no surprise, and therefore nothing new to learn. When something surprises us, that surprise signal (also called an error signal) can increase the change in connection strengths. The brain has different surprise signals, one of which is *dopamine*. We usually think of dopamine as associated with reward or pleasure, but dopamine is not just about the experience of pleasure—rather, it helps us learn to do what gets us reward. Dopamine is not released when a reward is received as expected and there is no surprise element. Instead, dopamine is released when *unexpected* reward happens, which in turn speeds up the processes of neuroplasticity. That causes the connections to neurons that are already firing together to increase even more. Scientist Shaowen Bao, working in Professor Mike Merzenich's neuroscience lab at the University of California, San Francisco, showed that when dopamine release was artificially stimulated in rats while they listened to a tone, the effect was so large that it actually increased the amount of their brain tissue that responded to the tone.

This process of neuroplasticity is how our beliefs change at the neural level. If we hear a bell, and then food appears, we are initially surprised. Quickly though, we learn that the bell means food is near. Even if we don't see the food right away, we now believe it will soon be there. Our brains have updated beliefs by changing the strength of connections between neurons, and this rewiring also allocates more neurons to carry information about this new belief. But, despite all the learning mechanisms, sometimes our brain gets off track and forms mistaken beliefs, making connections that seem appropriate but are actually based on misperceptions.

Mistaken Beliefs and How We Update Them

In the mid-2000s, the US Department of Defense had a problem. The US military had just invaded Iraq and toppled Saddam Hussein's regime. In the lead-up to the Iraq War, the US had intelligence agencies working with classified information trying to determine whether Iraq was developing weapons of mass destruction. The intelligence community saw indications that Iraq was developing chemical and biological weapons and importing the raw materials for nuclear weapons, like uranium ore and aluminum tubes for enriching the ore. If Iraq was allowed to develop nuclear weapons, this could pose a severe threat to US interests. In 2003, President George W. Bush acted on resulting intelligence reports by invading Iraq at a cost of one trillion dollars, thousands of American lives, and hundreds of thousands of Iraqi lives.

After the invasion, the US found almost no nuclear, biological, or chemical weapons in Iraq, at least nothing like the intelligence reports had suggested. This led to a fair amount of soul-searching in the intelligence community—how could they have gotten it so wrong? The Intelligence Advanced Research Projects Agency (IARPA), part of the US Office of the Director of National Intelligence, was tasked with funding research into how intelligence analysts make mistakes and form inaccurate beliefs about some part of the world, and how to help them avoid future mistakes. Earlier in my career, I had occasionally done research funded by the CIA, Office of Naval Research, and the Air Force Office of Scientific Research. This time, I was called on to work as part of the IARPA-funded research team with Raytheon, a large defense contractor, which serendipitously led to our findings described above about how the brain learns from surprise. Working with IARPA gave me a firsthand, though unclassified, look into how the government makes sense of sensitive data and tries to stay one step ahead of other countries' intelligence communities.

Intelligence analysts, those government employees charged with combing through classified data and making sense of it, are also humans. Like all of us, they have cognitive biases that can sometimes lead to mistaken beliefs. Our cognitive biases are a double-edged sword. "Bias" is not necessarily a bad thing—it simply means that we make use of shortcuts when we form beliefs about something. But there is a trade-off between making judgments quickly versus making them accurately. Those shortcuts help us make mostly accurate judgments without taking too much time or effort, but they can also result in costly mistakes.

Let's look at how we make judgments in the first place. One common way we come to a judgment reflects what is called the *availability heuristic*. We often judge how likely something is by how easily we can think of examples of that thing happening. If we get a garden-variety flu, we may feel horrible for twenty-four to forty-eight hours, but we also probably feel certain it will pass based on previous experiences and the availability of past examples of recovery that easily come to mind. This is the availability heuristic, and it biases our probability estimates by basing them on how easily we can think of relevant experiences.

Most of the time, the availability heuristic is a useful shortcut. But because our ease in thinking of the relevant examples depends on our past experiences, the availability heuristic can also lead to misjudgments. If you had to guess, which would you say causes more deaths in the United States: chronic lower respiratory disease or Alzheimer's disease? Think about that for a moment. How many times have you heard about people dying from chronic lower respiratory disease? How often have you seen it in the news? If you're like most people, you might have a hard time thinking of specific instances. On the other hand, how many times have you heard about someone dying from Alzheimer's disease? Do you know someone who succumbed to it? Most people can think of several instances, perhaps even people they know. It turns out that if you live in the US, you are statistically more likely to die from chronic lower respiratory diseases than from Alzheimer's (33.4 deaths per 100,000 versus 27.7 per 100,000 in 2023). Does that seem surprising? If so, you are not alone.

Psychology research has shown that people are generally bad at estimating probabilities and tend to use shortcuts, or heuristics, which can often provide useful estimates but other times get it seriously wrong. The availability heuristic is just one way we estimate the probability of something happening. We don't know offhand how likely something is, but we try to get a decent estimate based on how available the outcome is in our memory. In other words, we see how easily we can think of examples of the thing happening. We might be able to think of people dying from Alzheimer's disease. If so, that means if we are asked how likely someone is to die from Alzheimer's, we might guess that it is more likely. Conversely, stories of people dying from chronic lower respiratory disease are arguably less likely to catch one's attention, so we might have a harder time thinking of examples, and it will seem less likely. The problem is that though we can often make a pretty good guess about how likely something is, sometimes the availability heuristic biases our estimates and causes us to err. Our estimates could be correct, of course, depending on the examples we personally are drawing on to make our guess (perhaps we *do* know a lot of people who have died from chronic lower respiratory disease). But if we don't have a lot of prior experience with the subject matter, our estimate of the probability of something happening might be inaccurate.

Another main contributing factor to the Iraq invasion of 2003 was *confirmation bias*, a common tendency that can lead us to form mistaken beliefs. Sometimes the problem isn't so much that we don't have the right examples available to us, or the necessary experience, but that we form an initial impression from the experience we do have, and then ignore subsequent evidence to the contrary. In this way, even experts like classified intelligence analysts can get probability estimates badly wrong. Once it *seemed* likely that Iraq was developing nuclear weapons, the analysts in question tended to look for further evidence of it and discount or ignore evidence that suggested otherwise. The intelligence community recognized a potential for confirmation bias even before the Iraq War. Four years earlier, veteran CIA analyst Richards Heuer published a book that includes detailed methods that can help analysts, and people from all walks of life, avoid confirmation biases.

Heuer's technique is called the Analysis of Competing Hypotheses, and we can use it to evaluate our own beliefs. *The basic idea is that we can overcome our cognitive biases by first recognizing them.* The first step is to list out the alternative stories. In the *Massachusetts v. Johnson* case discussed earlier, there are several possible narratives: First, Johnson committed premeditated murder; second, Johnson didn't plan to kill Caldwell but did so in the heat of passion; or, third, Johnson was merely protecting himself and acted in justifiable self-defense. The next step is to list all versions of the story along with all the pieces of evidence and evaluate each possible combination. Is the evidence consistent with the story? Or inconsistent? Or does it have no bearing on whether the story could be true?

In 1964, University of Chicago professor of physics John Platt wrote a classic article in which he argued that *instead of trying to prove our theory is correct, as scientists, we should focus our effort on trying to disprove theories.* Doing so, he argued, more effectively narrows down the range of possible stories until we find the truth. Heuer later argued a similar point, that evidence that is inconsistent with a story is the most useful evidence, and the best thing we can do is to look for *dis*confirmation of potential stories: try to disprove them. The story that remains most viable afterward is most likely to be true. In *Massachusetts v. Johnson*, even if most of the evidence is consistent with a particular story, we should put more weight on any evidence that is inconsistent. If we think Johnson was innocent, we should look for evidence that he actively lunged toward Caldwell with the knife, as that could disprove the story that Johnson was merely defending himself.

The question of whether evidence is consistent with a possible story can itself be difficult to evaluate. If we can think of examples of people carrying knives in order to commit premeditated murder, we might judge the fact that Johnson was carrying a knife as making it more likely that he was planning to kill Caldwell. This is the availability heuristic, and the first step to overcoming it is to recognize it. Beyond that, Heuer's emphasis on looking for disconfirming evidence can also mitigate our tendency to overestimate the likelihood of the truth of a story.

These principles of belief formation, including cognitive biases and ways to overcome them, apply to spiritual beliefs. On the one hand, we'll see evidence in part 2 that spiritual beliefs and practices we develop can improve health. In that sense, a belief that certain religious practices can help us become healthier is empirically true. Robert Wright, in his book *Why Buddhism Is True*, makes a similar argument in the sense that the practices Buddhism entails may be good for our well-being. Yet belief that helps us become healthier is not necessarily the same as empirically true.

Suppose I believed that eating fresh vegetables blessed by a holy person makes me healthier because the vegetables carry divine energy. I could be right that eating vegetables improves my health but wrong about why I believe they do. I would be committing a logical error known as *affirming the consequent*. Just because one event if true (blessed vegetables having hypothetical divine energy bring health) would imply another event (eating blessed vegetables makes me healthy), that doesn't mean that if eating blessed vegetables makes me healthy, then the vegetables must have divine energy. Scientific methods can help us guard against logical errors like this, for example, by testing whether vegetables that *haven't* been blessed by a holy person still make us healthy. If so, that would disprove my belief that the vegetables must be blessed by the holy person or else they won't improve my health.

Ideally, we would hold beliefs that are both empirically true *and* good for our health. We must also recognize that not all beliefs, including certain religious and spiritual beliefs, are good for our well-being. As Professor of Psychology Kenneth Pargament at Bowling Green State University and psychiatrist Dr. Samuel Weber have shown, interpreting sickness as divine punishment can result in spiritual distress. Perceived conflict between medical advice and religious requirements may lead patients to delay or avoid beneficial medical treatments. These sorts of beliefs can lead to poorer mental and physical health outcomes.

By understanding how we form beliefs, how we can be mistaken, and by learning some new methods for investigating whether our beliefs are true, we can give our own beliefs a checkup. We may become more successful in arriving at beliefs that are both empirically true and good for our health. As we will see in part 2, we can learn to cultivate spiritual beliefs with the potential to improve our mental and physical health.

PART II

How Beliefs Affect Our Health

4

How Beliefs Affect Mental Health

IN THE WAKE OF MY BRAIN TUMOR DIAGNOSIS, ONE SUNDAY morning a friend at the church I attended mentioned a local group of Christians from several different churches who believed that prayer could heal. As the outlook for my condition was bleak with no real medical hope of a cure, I decided to visit and see if their prayers could help me. The next day after dinner, Candy and I drove across town, arriving ninety minutes before the scheduled meeting, because the group's leader, James, had offered in-person prayer at that time.

I walked in expecting soaring music, throngs of people, clean white walls, and lofted ceilings. Instead, I found a circle of half a dozen plastic folding chairs in the side room of a church building in obvious need of repair. The scene was underwhelming compared to the fantasy I'd constructed. I took stock of James—white, middle-aged, casually dressed in loose-fitting blue jeans and a faded long-sleeve shirt—finding no signal of anything out of the ordinary. After introductions, Candy and I joined the circle, and the group began to pray for my healing.

I was accustomed to prayers like "Dear Lord, if it be Thy will, please heal so-and-so of their disease. Amen." By contrast, James and the others gently but firmly spoke to the disease itself, to the effect of, "By the authority of Jesus, I command the tumor to shrivel up and leave Josh." Their commands took me aback somewhat. It shocked me that they thought they could just tell a tumor to disappear. James and his group clearly not only hoped that God would answer—their language and demeanor indicated that they expected results. Moreover, they seemed to see supernatural intervention as such a common and likely occurrence that their first response to my illness was to begin praying rather than commiserate or suggest medical interventions or special diets. I was tempted to excuse myself and go on my way, but the scientist in me asked myself: Why did they seem to expect healing to occur through their prayers, and even assuming that prayers did have some positive effect, how would that work?

What I experienced at that prayer meeting did not fit my previous cognitive map. I had always been religious. I'd attended church since childhood and believed in the existence of God. But I didn't expect God to act beyond what science predicted. I'd mostly adhered to a naturalistic view of the world, adopting *methodological naturalism* as a framework for scientific research. I did my scientific experiments and interpreted them on the assumption that the world is and should be understandable through the study of natural processes, not by looking for supernatural disruptions (even if the supernatural exists).

But not everyone sees the world that way. On the one hand, there are those who believe in *absolute naturalism*, which categorically excludes the possibility of any supernatural explanation. As science fiction writer Arthur C. Clarke put it, "Any sufficiently advanced technology is indistinguishable from magic." Or to quote Marvel's film *Thor* (2011): "Your ancestors called it magic, but you call it science. I come from a land where they are one and the same." By this reasoning, if we can't find a naturalistic explanation, it's because we don't yet understand the phenomenon. The supernatural exists only as a "God of the gaps" in our scientific knowledge. As science advances and provides new explanations for how remarkable phenomena can be explained by freshly discovered natural processes, supernatural explanations fall, leaving an ever-shrinking role for God's activity in the universe. On the other hand, there are those who believe that the supernatural is commonplace or that a God who created natural processes can supernaturally intervene at any time. Belief in the supernatural is shared by many religions, from Judaism's teaching that Yahweh parted the Red Sea for the Israelites to escape an Egyptian army to Sanātana Dharma's (Hindu) tradition that water from the Ganges has healing properties because this river is a physical manifestation of the goddess Ganga. For many spiritual individuals, God or some higher power is active in many or even all observable phenomena.

Methodological naturalism offers some space for a middle ground. We might not absolutely rule out the supernatural but view it as so unlikely as to be mostly irrelevant. My own methodological naturalism meant that I saw healing as something that occurred because of a chain of natural processes, including human intervention, but was not supernatural. By this logic, any claim that God must have caused healing supernaturally requires, at minimum, that we exhaust all possible natural explanations.

After one and a half hours of personalized attention, James said it was time to start the official prayer meeting, but three group members offered to continue praying for me. I appreciated their concern and accepted, despite the awkwardness. They walked us down a dimly lit hallway with peeling plaster walls to what looked like a Sunday school classroom for young children, with child-size seats, toys, and wall decorations to match. "In the name of Jesus, I command the spirits of cancer to leave Josh," a woman from Ghana, named Agrippina, intoned sternly as Candy sat quietly beside me. They paused periodically to ask questions about my family and spiritual background and to check how I was feeling now and if I noticed any effects of the prayers. They continued to pray with me like that for another hour and a half.

It was around this point that a change started to come over me. Early in the prayer session, I'd felt anxious, awkward, and skeptical from the weirdness of it all—the harsh lights, their jarring commands, my discomfort at revealing my personal details and death diagnosis to strangers. But after three hours, an ineffable sense of peace cascaded down. It was as if all my anxiety had fled and been replaced by utter calm and relief. A brief engagement with counseling in the past had produced a similar but less intense effect, and I had had some deep experiences with prayer several years before. But my encounter with the active prayer of James's group seemed much more—for lack of a better word—*powerful*. Their spiritual practices had lifted a huge weight off me. Despite only having just met this group, their willingness to spend so much time suggested that they cared for me and my possible recovery, or at least that they found the experience of offering prayers for healing to be fulfilling. Their hope and confidence were infectious. I sensed profound love, not just from the people who'd prayed with me but directly from God. When the session ended, I left with renewed hope.

How did their prayers have such a positive effect on my sense of well-being? I certainly *wanted* the prayers to heal me. But I was also intrigued as a methodological naturalist looking for the neuroscientific basis of what was happening. I *sensed that* the prayers were helping. But how might I scientifically explain *how* the prayers were helping? Could our discussion alone, which had seemed vaguely therapeutic, be responsible for my improved mental state? Could there also be effects on my health that would go beyond my mental state?

Reframing

David Washburn, Professor of Psychology at Georgia State University, and his colleagues studied how prayer can improve our mental state. Their study asked individuals (specifically college students) to pray about a concern in their life for ten minutes (exam, relationships, finances, or major decisions). Another group of students were instead asked to contemplate or think about the situation in their life, but without praying. Afterward, they were given difficult cognitive tasks, such as finding words hidden in a grid of letters, that required them to respond quickly. The researchers found that, first, the students who had prayed were able to respond faster to the challenging tasks than those who had not prayed. The praying students had unburdened themselves by asking God's help in resolving life concerns. This freed up cognitive resources so that they could focus their attention and perform better. Second, the students who prayed were better at finding hidden words related to the concerns they prayed about. If they had prayed about their finances, they were better able to find words like "money" hidden in the letter grid. Their cognitive processes shifted from worry to problem-solving, allowing them to find relevant information more easily. The effects were strongest in those who regularly engaged in religious activities such as prayer.

Prayer meetings like the one I attended, especially when prayers include concerns about mental or emotional health, often resemble secular counseling. The prayers provide perspective on, or *reframe*, the trauma involved. Similarly, modern psychotherapeutic approaches, such as cognitive behavioral therapy (CBT), often suggest using cognitive reframing to change how we think and feel about stressful situations. For example, we may experience a traumatic event, like a sudden illness or accident. Our subsequent attempts to process what's happening may lead us to an empirically untrue or dysfunctional belief. We may conclude that we are somehow to blame for an illness or accident, or that it is our fate to suffer in life because there's something wrong with us. None of this is true.

Reframing alleviates stress by changing the lens through which we view something. In many cases, the answer to needless suffering is to reframe positively—and change our beliefs to reflect things as they really are. In the example above, we would reframe the false belief that there's something wrong with us to an understanding that illnesses and accidents happen to everyone. Similarly, rather than believing that it is our fate in life to suffer, we might reframe to see ourselves as resilient for surviving the traumatic event. By reframing, we can replace catastrophizing or all-or-nothing thinking with beliefs about ourselves that might reflect our capabilities and a more positive lens on the situation. Reframing isn't about falsely asserting the positive. The goal is to stop engaging in negative thinking—and to reflect the truth.

Cognitive reframing has been shown to be highly effective. In this process, typically, a therapist guides a patient through several steps to help challenge and replace negative thoughts. The first step is to increase awareness of our thoughts and identify the faults in them. Then, in step two, the therapist uses the Socratic method to challenge irrational, illogical, or harmful thinking errors. As part of this questioning, we identify a thought we feel needs examination, consider the evidence for and against this thought, and make a judgment on it, including whether it is a black-and-white situation or there is room for shades of gray.

Making a judgment based on a thought may feel very different from the standard mindfulness advice to gain nonjudgmental distance from our thoughts and emotions. It's true that this is a different approach, but many people find it useful as it goes more directly to understanding the belief that is causing the negative emotion. It is also compatible with mindfulness. There is a mindfulness-based variant of CBT in which we also learn to accept unpleasant emotions and then replace them with calmness or compassion. By determining the validity of a belief, CBT frees us to eliminate false beliefs, thereby freeing us from the accompanying negative emotions. The third step is to figure out how to respond in a way that matches the truth. This step might include the use of guided imagery, focusing on feelings we're experiencing, and, finally, reframing. This often involves prompted dialogue, a thought journal, questions about "what if," and *de-catastrophizing*, or viewing the difficult situation as not catastrophic. *Overall, the goal is to reframe or transform the negative thought about the situation into a more positive and constructive way to proceed.*

While we may experience cognitive reframing in secular settings like a counseling office or religious settings like a prayer meeting, both can be effective at shifting our beliefs in a positive direction. A study in 2015 randomized people to receive either standard CBT or a religiously framed variant of CBT that invoked religious themes and parts of sacred texts from the participants' own traditions, with versions customized for Christian, Hindu, Muslim, Buddhist, and Jewish participants. The hypothesis was that religious framing would help participants engage with therapy more effectively as they saw connections between CBT principles and their faith traditions. The results showed that religiously framed CBT was at least as effective as secular CBT.

Cognitive reframing isn't confined to the psychologist's office. It can be used in other forms of therapy as well. If an accident has left us unable to do certain things, a physical therapist might help us let go of unrealistic goals for rehabilitation and focus on manageable next steps. This can bring satisfaction for achieving small goals rather than disappointment at not achieving larger objectives. By letting go of unrealistic standards we put upon ourselves, we might discover new inner strength and resilience.

How Healing Prayer Changes the Brain

Prayer interventions, such as the one I experienced under James's leadership, have direct effects on the brain as well as on our emotions. In the mid-2000s, Harold Koenig, an MD and professor at Duke University and director of its Center for Spirituality, Theology, and Health, worked with Peter Boelens to study in-person prayer for healing. Boelens, also an MD, listed his affiliations as "University of Mississippi and Shalom Prayer Ministry," echoing the kind of science-faith research partnership seen in the Dalai Lama's partnership with Richard Davidson and Jon Kabat-Zinn. Koenig and Boelens's question was twofold: Could in-person prayers for healing help improve mental health, especially anxiety and depression? And if so, how might the prayers affect the brains of prayer recipients?

Boelens recruited adults who had been diagnosed with depression, and most also had anxiety. Half were randomly assigned to receive prayers, and the other half received none. Every week for six weeks, each participant in the prayer intervention group received prayer, one-on-one, in person, in a Christian tradition for healing, for sixty to ninety minutes. During this time, the person offering prayers used a mix of spontaneous and scripted prayers, focusing on specific stressors in the participant's life and healing of childhood emotional traumas and traumatic memories. The prayer recipient just sat quietly and listened.

At the end of the six weeks, all participants were reassessed for anxiety and depression, and then reassessed again one month later. Their cortisol levels were also measured before and after the intervention period, as biological markers of stress. After six weeks, those who received prayer reported significantly lower levels of depression and anxiety, while those in the control group had unchanged depression and anxiety levels. Despite all this, the researchers found no change in the participants' cortisol levels. Their mental health had improved, but their biological markers of stress did not. One year later, Boelens followed up to see how the depression and anxiety levels looked. Did the effects of prayer last that long? The answer was yes—the participants showed a reduction in anxiety and depression following prayer, and that improvement persisted for at least a year.

There was still a question of what was happening in subjects' brains as their mental health improved. A few years later, the same research group repeated the experiment, but this time asking whether in addition to improving anxiety and depression symptoms, the prayers might cause changes in the brain. They replicated the finding that the depression and anxiety symptoms dropped significantly after prayer. To examine brain changes, the team used an fMRI scanner to measure brain activity before the six weeks of prayer, and again after. While in the scanner, participants were asked to think about bad memories and feelings, then later think about neutral memories or feelings. Comparing before and after prayer, sensitivity of the medial prefrontal cortex to bad memories increased significantly. Moreover, the more the depression scores decreased, the more the prefrontal cortex activation for the bad memories also lessened. It was as if the brain was being rewired by the prayer experience so that it could respond properly, and that restoration corresponded with reduced symptoms of both depression and anxiety.

How Beliefs Change the Brain

Whether or not people pray about the causes of their anxiety, experiences that induce anxiety often push people toward religious beliefs. There's an old saying that "there's no such thing as an atheist in a foxhole." In other words, someone in distress, such as a soldier in a foxhole, often looks to religion for comfort. This can be seen in groups, not just individuals, and even on a national scale. On the evening of September 11, 2001, following the terrorist attacks that destroyed the Twin Towers in Lower Manhattan and part of the

Pentagon, the US Congress gathered as a group on the Capitol steps and sang "God Bless America" together. It was a remarkable display of bipartisan unity, also notable for the overt religious theme of calling on God for help, which Congress was not otherwise known for. The song itself was written during World War I and found renewed popularity in World War II.

This kind of faith response in the throes of uncertainty or danger is not unusual. As Michael Inzlicht, professor of psychology at the University of Toronto, notes, "political instability is often followed by increases in faith." When times are uncertain, people may turn to religious faith in part to help relieve anxiety. Research suggests that this strategy can be effective.

Late in the first decade of the twenty-first century, Inzlicht and colleagues carried out a series of experiments looking at how beliefs can affect the brain and in turn improve anxiety. Their basic idea was that people are motivated to make sense of their environment, and religious belief provides a coherent story, which in turn reduces anxiety. To test this hypothesis, Inzlicht and his colleagues focused on the anterior cingulate cortex, which, as we learned about in previous chapters, shows increased activity when there is some threat or problem that requires action to avoid or resolve.

The first experiment required manipulation of neural activity in the anterior cingulate. To achieve this, Inzlicht induced a state of internal conflict by asking subjects to perform something called the Stroop task (named after the researcher, John Ridley Stroop, who created it). Subjects are shown a single word at a time—the name of a color, such as "red" or "blue." The word is also written with a particular ink color. Sometimes, the word "red" is written with red ink, and sometimes with blue. Then the subjects are asked to, as quickly as possible, name the *ink* color of the word while ignoring the word itself. If shown the word "red" written in blue ink, the correct answer would be "blue." People find this task difficult because we typically read a word and ignore the color of the letters that make up the word. Experiment subjects often get the answer wrong, reading the word and blurting out "red," which is incorrect. After a few trials, people realize the nature of the task and can perform it well, even though it remains difficult.

The basic idea behind the Stroop task is that it creates a state of conflict in our brains. We are torn between reading the word and naming the ink color. That conflict leads to more activation of the anterior cingulate cortex, which in turn may help us resolve the conflict and generate an appropriate response. Moreover, mistakes (such as reading the word instead of identifying the color) generate strong activation in the anterior cingulate cortex, as does conflict even if we don't make a mistake. The anterior cingulate is always vigilant, looking for potential problems even before they happen and helping us to avoid them. The Stroop task thus served its purpose as an experimental technique that can change anterior cingulate activation. We'll see in a moment how religious belief works to reduce anterior cingulate activity, and how that reduces our anxiety. But first let's look more closely at the basics of how the anterior cingulate cortex contributes to our anxiety.

The anterior cingulate is a key player not only in anxiety but also in obsessive-compulsive disorder (OCD). People with OCD generally have an overactive anterior cingulate, and in the event that all other treatment options fail, a neurosurgical treatment that removes part of the anterior cingulate can effectively reduce OCD symptoms. Based on experimental findings with OCD, interventions that reduce anterior cingulate activity should also help resolve anxiety related to a feeling that something is wrong and needs to be fixed or avoided. In 2013, Josef Parvizi and colleagues at Stanford tested this idea, except in reverse—they wanted to see what happens when they artificially increased anterior cingulate activity by electrically stimulating it. They had a unique opportunity to do so, because they had two patients who had electrodes implanted in their anterior cingulate regions in preparation for surgery to treat epilepsy. The researchers stood by the patients' beds, connected a small electrical stimulator to the electrode wires that passed through their bandaged heads into their brains, and started passing electrical current to stimulate the anterior cingulate region. As soon as the electrical current started flowing, the first patient began to show an increased heart rate and said, "I started getting this feeling like . . . I was driving into a storm. That's the kind of feeling I got. Like, almost like you're headed towards a storm that's on the other side, maybe a couple of miles away, and you've got to get across the hill and all of a sudden you're sitting there going 'how am I going to get over that, through that?' And that's the way my brain started functioning."

It was as if electrically stimulating the anterior cingulate instantly turned on the patient's anxiety, and shutting off the stimulation likewise instantly turned off the anxiety. This finding suggests that if we can reduce anterior cingulate activity, then maybe we can directly alleviate anxiety.

Let's return to the question of how prayer can reduce anterior cingulate cortex activity and how that may reduce anxiety. To investigate this, Inzlicht recruited participants from diverse religious backgrounds, including atheists. In their first study, the researchers asked participants to rate how strongly they held religious convictions. Those who had stronger religious convictions showed weaker activation in the anterior cingulate when they made a mistake in the Stroop task. In a second study, Inzlicht asked participants a single, slightly different question, about how much they believed in God. Responses ranged from "certain God does *not* exist" to "certain that God exists." Again, those who believed in God more strongly displayed smaller cingulate activation when they made a mistake. Also remarkably, the researchers found that those with stronger religious beliefs in general and belief in God specifically made *fewer* mistakes in the Stroop task. Inzlicht interpreted this to mean that religious beliefs make the brain less sensitive to anxiety from conflicting situations or mistakes, which helped religious subjects perform better.

Still, this leaves open a question of causation. Does religion make the brain less sensitive to conflict or mistakes, or does less sensitivity to potential mistakes make people more likely to hold religious beliefs? To address this question, Inzlicht did another experiment, where just before their brain scans, subjects had to write a short paragraph for five minutes. Half of them were asked to write about what religious beliefs meant to them personally, while the other half was asked to write about what their favorite time of the year meant to them personally. The idea was to increase awareness of religious beliefs in the participants who wrote about religion, thus increasing any effects by which religious beliefs might influence anterior cingulate activation. The group that wrote about religious beliefs subsequently had less anterior cingulate activation when they made mistakes. This suggests that focusing attention on religious beliefs can reduce distress and anxiety, specifically by reducing the sensitivity of the anterior cingulate cortex. In other words, *religious beliefs here are the cause of reduced anxiety, not a result of it.*

Even when we're not consciously reflecting on our religious beliefs, their continued presence in the back of our minds can also help reduce anxiety. In another study, Inzlicht and colleagues recruited both religious people (whom they labeled theists) and those with low religiosity (whom they likewise labeled atheists), and asked them to do an apparently unrelated task of unscrambling words of a sentence before doing the Stroop task. Half the subjects were asked to unscramble words related to religion, such as "prophet," while the other half were asked to unscramble words that had nothing to do with religion. In total, there were four groups: theists who unscrambled religious-themed words, atheists who unscrambled religious-themed words, and theists and atheists who each unscrambled nonreligious-themed words. The idea was that unscrambling words with religious themes would help prime subjects' minds to focus on religious beliefs. The experiment worked—among the theists, unscrambling religious-themed words led to reduced anterior cingulate activation for mistakes. The religious-themed words did not help the atheists, though—they showed more, not less, brain activation for mistakes when they had just unscrambled the religious words. The other two groups of theists and atheists who unscrambled words that were not related to religion showed no difference between the groups. Their brains activated at the same level, regardless of whether they were theists or atheists, when they made a mistake. This finding shows that simply unscrambling nonreligious-themed words does not affect theists and atheists differently. Instead, the reduced error-related brain activation in theists was specifically driven by the religious themes in the words they unscrambled.

How exactly religious beliefs could reduce anterior cingulate activity was the next question Inzlicht considered. Are the people who focus on their religious beliefs simply less sensitive to their mistakes, perhaps not realizing they have made one? Or are they less motivated to do the Stroop task? Perhaps they are aware of making mistakes but feel less bothered by them? It turns out that the more religious participants performed the Stroop task *better*, not worse than other participants. Inzlicht concludes: "Religious people are less anxious and defensive about the errors they make." In support of this claim, the religious participants didn't startle as much in response to a loud noise. They were both more careful in performing the Stroop task and less bothered by mistakes they made in the process. By helping make sense of the world and providing a sense of order, meaning, and purpose (as well as by providing a social network of like-minded believers), religious beliefs help reduce anxiety and allow us to perform better. *Whatever our specific religious beliefs may be, affirming our beliefs, and the order and meaning they provide for our place in the world, may help reduce both our anterior cingulate activity and our anxiety while also improving our ability to perform mentally challenging tasks.*

Addiction

In addition to experiencing anxiety, many of us have struggled with addiction or lost loved ones to addiction. My uncle passed from liver failure after a lifelong battle with alcohol dependence. He was just fifty-seven years old. Addiction is a terrible disease, but religious beliefs can change our brains to help overcome it. Around 2007, I became interested in studying addiction to answer a larger question of why people choose risky behaviors that otherwise are not in their best interest. A core feature of addiction is that people drink and use substances despite the adverse effects on their lives.

In 2009, my colleague Peter Finn, a professor at Indiana University and a longtime addiction researcher, and a new graduate student in my lab, Sarah Forster, set out to discover which brain mechanisms can help people recover from addiction. Sarah heroically recruited and performed fMRI scans on a number of people with alcohol dependence who had just entered treatment. Our participants were not the easiest to work with; their new abstinence from drinking left them irritable, but Sarah's calm demeanor helped defuse a few tense situations. When the results were in, we found that when the subjects played a gambling task in the fMRI scanner, some of them showed very strong activation in the amygdala when they lost a gamble, while others showed weaker activation. We kept in touch with both groups for the next three months. We found that the ones with stronger amygdala activation when making a mistake were more likely to fail out of their addiction treatment, while the ones with weaker amygdala activation were more likely to be successful.

Just by measuring subjects' amygdala activation, *we could predict with as much accuracy as any other method how successful their addiction treatment would be.* This makes sense, because stress and its accompanying negative emotions can activate the amygdala, and increased stress is a common factor in addiction relapse. At the end of their three-month treatment period, we scanned them again. These later scans found that in subjects who successfully abstained from alcohol, the cingulate became more sensitive to the risk of the gamble, while the prefrontal cortex became less sensitive. Their brains had changed as they overcame their addiction, and the normal ability of their brains to respond to risk was restored.

Around the same time, Peter, Sarah, and I visited a faith-based recovery program just outside Bloomington, Indiana. The group was seeing remarkable results of people overcoming their addictions by adhering to a faith-based regimen of prayer and other Christian spiritual practices. While we didn't study the program further, it raised the question: Was prayer helping participants overcome their addiction? This was a reasonable question, given that Buddhist meditation has been shown to reduce amygdala reactivity.

In 2017, Professor of Psychiatry Marc Galanter and colleagues at the New York University School of Medicine addressed the question of whether prayer works like meditation to reduce amygdala reactivity. Galanter had previously found

Alcoholics Anonymous (AA) to be effective in helping individuals abstain from alcohol. Galanter next studied the effect of AA prayers on brain activity and alcohol cravings. The AA twelve-step recovery program includes prayer addressed to "God" or a "Higher Power"—admitting to being powerless against addiction, calling on God to help, and so on. While Galanter's subjects were in the fMRI scanner, they were shown images related to alcohol, designed to induce cravings. Before that, the participants were instructed to do one of three tasks: read the AA prayers and remember them, read a random article from the news and remember it, or else simply look at the forthcoming alcohol images passively. The researchers found that when people read the AA prayers, the subsequent alcohol-related images were processed differently in their brains. One of the brain regions showing more activation was the prefrontal cortex, which is associated with attention and self-control. When subjects' brains showed more activation in the prefrontal cortex, they reported reduced alcohol cravings. The prayers seemed to help their brains overcome their cravings. *If we can increase our prefrontal cortex activity, for instance, by using prayers like those in AA, we, too, may become more successful in overcoming addictions.*

All this suggests that religious beliefs can change our brains and, in the process, improve our mental health. By focusing upon and affirming our beliefs, we might experience reduced mental health symptoms like depression and anxiety or overcome addictions. At the same time, we might even be able to rewire our brains to process threats and uncertainty in ways that make it easier for us to function in daily life. None of these findings proves or disproves that God actually exists or answers our prayers—but the data do suggest that our religious beliefs and spiritual practices such as prayer may help us feel better. Our beliefs are not only important for overcoming mental health challenges. As we'll see next, our beliefs can also improve our overall sense of well-being.

5

How Beliefs Affect Well-Being

I first met my wife, Candy, in 1996, less than a month after I had moved to Boston from San Diego to start graduate studies for my PhD. Although I knew no one in Boston when I arrived, I soon found a new Vineyard church meeting in the living room of a recently renovated apartment in East Cambridge, in one of the many three-story apartment buildings there commonly known as triple-deckers. Of the twenty or so members, half had recently graduated from Harvard, where they had belonged to an Intervarsity Christian Fellowship chapter, the same Christian fellowship organization I had joined as an undergraduate in San Diego. The other half, like me, were California transplants. Several were Stanford grads from a Vineyard church in the Bay Area that was part of the same national Association of Vineyard Churches as the one I had attended in San Diego. With so many experiences in common, the group felt remarkably familiar, even though I had only just met them all.

One of those church members was Candy. She had also just returned to Boston to start her PhD and joined the church with friends from her Harvard undergraduate days. Her smile immediately caught my attention. Through meetings at the East Cambridge house, I got to know her, and she struck me as someone special from the outset. She was as fascinating and full of fun as I had predicted, and we were soon dating. We had a surprising amount in common: We both came from families that hadn't had a lot of money when we were growing up. I had lived in a trailer park until I was twelve, and she had helped her family recycle aluminum cans to buy groceries. We both grew up in California. And, of course, we both had our faith. We even had similar church backgrounds: My family had been Baptist, hers Mennonite Brethren. Mennonites were a lot like Baptists, but pacifist. More than that, we enjoyed discussing all kinds of intellectual topics. We were definitely a match. Soon, we talked about marriage.

Paradoxically, the more I thought about getting married, the more anxious I became. Memories of how my parents had struggled for years with their marriage and divorced just four years earlier had changed my beliefs about how marriages work and focused my attention on how likely they are to fail. My worry had recently increased when the last remaining marriage in my extended family also ended in divorce. I loved Candy and thought that, objectively, we could build a lasting marriage. But drawing on memories of my family, my brain predicted that mine would end in pain and divorce.

The issue came to a head one day as I secretly decided that I was going to propose to Candy. I went out to buy a ring. Instead, on the subway train from Cambridge to the diamond district in Boston, my anxiety about getting married skyrocketed.

I felt trapped. I couldn't imagine losing Candy, and yet I was terrified of marriage. I began to have thoughts of harming myself as the only way out. I went back home without buying a ring.

After that, I met with some of my friends from the Vineyard church and confided my anxiety. They gathered around me, listened, prayed with me, and offered support when I was at my lowest. A few days later, I attended a weekend conference on healing emotional pain at another Vineyard church an hour away. The speakers talked about how God heals emotions and prayed for me to receive healing. It was then that I realized just how painful my parents' divorce had been for me. Although I had moved as far across the country as possible, thinking I could simply avoid reminders of a broken home, I couldn't outrun my own pain and my own cognitive map. I also realized that my gloomy beliefs about how marriages are likely to fail were just that—beliefs—and they were coming from seeing the failed marriages in my own family. I was not destined to repeat the same patterns. My marriage with Candy could work. With that, I reframed my beliefs about marriage. I left that conference with a smile, looking like a different person. Some months later, in one of the happiest moments of my life, I proposed to Candy, who said yes, and we married the following year, in 1998.

Social Support

With support from friends in my church, and grounded in new beliefs about how marriage can work well, I got through an emotional low point. My experience is one of countless examples of how social support can make the difference between life and death. Social support can improve our well-being and even lower our cortisol, as shown by Professor Dr. Markus Heinrichs, now Chair of Biological Psychology, Clinical Psychology, and Psychotherapy at the University of Freiburg in Germany. Heinrichs tested a group of participants in a laboratory setting with a social stress test, in which the research subjects had to give a public speech and solve difficult math problems in their heads on the spot while an audience looked on. For most people, this would be at least mildly stressful. Unsurprisingly, subjects' cortisol levels rose measurably as they did the stress test. The twist is that half the subjects were asked to bring a good friend with them to provide support as they did the stress test. The other half of the group came alone. The results showed that those who came alone had higher cortisol levels after the stress test than those who came with a friend. As we saw earlier, beyond the benefits of lower stress on our mental state, keeping cortisol levels under control can support overall health.

Social capital, or our network of friends and family whom we can call on for support, enhances individual well-being. In the animal kingdom, cooperative behavior protects against threats and aids reproduction. Ostriches, who have a great sense of smell but poor eyesight, cooperate with zebras, who possess the opposite strengths, to stay alert to predators. Male dolphins work together to guard fertile females. Pied babbler birds care for a dominant breeding pair's offspring so that the dominant pair can have a second brood in one season. We saw in chapter 3 that social cooperation in humans, as with animals, can increase our ability to survive and reproduce. Unlike animals, humans may no longer require high levels of cooperation to survive. But we still seek out and benefit from social connections because they enhance our well-being. Our brains are wired for social interaction and emotional connection.

Neuroscience helps to explain why we need social support by showing a direct correlation between social activity and brain activity. Social contact enhances brain development in people of all ages, and adolescents in particular. Social interaction, especially touch, causes the pituitary region of our brains to release oxytocin. Sometimes called the love hormone, oxytocin may in turn improve our ability to enjoy positive social interactions, helping us recognize and respond to emotions in other people. Moreover, oxytocin helps reduce our cortisol levels when we're stressed, thus improving our health, as we saw in chapter 2. Other studies bear out this relationship between social interaction and health. In one of the all-time longest-running scientific studies of human behavior, Professor of Psychiatry Richard Waldinger and colleagues at Harvard studied human flourishing over eighty years and multiple generations, interviewing families regularly over the course of their lives to find out what led to fulfillment and happiness. The study continues to follow a few of the original participants who remain alive as well as their children, but the results to date show that social interactions matter—our lives are longer and happier largely because of our social interactions in all their various forms.

Spiritual beliefs can help meet social needs by facilitating positive social interactions. Many religious traditions encourage attendance at a church or similar religious space to participate as a group in worship, teaching, and other community activities. Even controlling for other factors like a healthy lifestyle, simply attending religious services regularly led to a 46 percent reduction in mortality among older adults. People who participate most often in organized religious activities such as prayer meetings, or nonreligious community organizations like a bowling league, on average report higher levels of happiness in addition to better overall health. There is literally a mood effect, whereby social interactions and the oxytocin release they support can provide relief from loneliness and/or increase happiness. More remarkably, some spiritual practices can provide us with the benefits of social interaction even without our interacting with other people.

A neuroimaging study by Uffe Schjoedt, Professor of Religion at Aarhus University in Denmark, sought to isolate the mood effects of prayer itself from the social benefits of praying with a religious community. In the experiment, highly religious participants were scanned with fMRI while saying scripted prayers from memory (the Lord's Prayer) or unscripted, spontaneous prayers. As a control condition, they also recited scripted nonreligious material (a nursery rhyme) or unscripted nonreligious speech (making wishes to Santa Claus). When subjects engaged in unscripted, heartfelt pleas to God compared with making wishes to Santa Claus, their brains showed increased activity in specific regions, especially the *temporoparietal junction*, *temporopolar region*, and *anterior medial prefrontal cortex*. These regions, especially the temporoparietal junction, are associated with interacting with another person and with *theory of mind*, which is the ability to understand what someone else is thinking. This suggests that not only were the subjects' brains processing prayers like they were talking to another person, but also that they did so with equal attention to how the other person would understand what they were saying. It was as if when praying, they were participating in a "relationship" with God. Recorded neural activity is consistent with a conclusion that subjects perceived themselves to be speaking to God directly, with God hearing their prayers.

These results do not prove that participants were actually interacting with God as one would with another person, nor do they prove God's existence. The study does suggest that subjects who used unscripted prayers might have subjectively experienced a direct connection to God that provided a sense of social connection similar to interacting with others at a church service or socializing with friends over a cup of coffee. That experience in turn led to a feeling of social support, and the overall well-being that entails.

Forgiveness

While positive social interactions can help our well-being, sometimes interactions become hurtful. When I was twelve years old, my father worked as an administrator for a small Christian college, led by a man who was both the president of the college and the pastor of the church we attended. In difficult times, church members often ask the pastor of their church for spiritual guidance. My parents were having marriage problems at the time, so my mother went to the pastor for help. Some months later, when it came time to renew my father's annual contract with the college, the president called him into his office.

"Bill, we're not going to renew your contract," the president told him. "We're letting you go."

Shocked, my father asked, "You're not? Why not? I've done my job well. No one has raised concerns about my performance."

"Well, it's not about performance. I understand you've been having marriage difficulties. This is a Christian college. We can't have someone in senior leadership who is having the kinds of domestic difficulties you have. It doesn't set a good Christian example for our students."

Still stunned, my father cleaned out his large wooden desk and said goodbye to his coworkers. He ended up on unemployment, and our family finances got tight. Not knowing where else to turn, my father asked the pastor, the same one who had terminated him, for spiritual guidance. The pastor declined, telling him to go elsewhere. Up until then, our experience of church was as a place where we could find social connection and support. My father's experience with the pastor felt like a betrayal. It was if my father had gone to a hospital for help, only to get shot at and then sent away. All of us in the family—my parents, my sister, and I—also felt the sting of this betrayal and stopped attending church altogether. Four years would pass before we attended another church, of a different denomination.

Meanwhile, some months after being terminated, my father sensed that he had a religious duty to forgive the pastor. He made an appointment with the pastor to extend forgiveness, but the pastor seemed to barely remember, much less care about what had happened, nor did he ask for my father's forgiveness. That didn't matter to my father. Extending forgiveness was for himself, not the pastor, because doing so freed my father and gave him the chance to move on. He soon found a similar job working for a friend at a different Christian college in another city.

Like my father, many have found that religious grounding helps them to forgive. In a 1998 study of 1,445 individuals in the United States, respondents were asked to rate how strongly two statements were true about them: "I have forgiven myself for the things I have done wrong," and "I have forgiven those who hurt me." Those who were affiliated with a religious tradition, especially since before age sixteen, were more likely to have traits of forgiveness later in life. The more an individual looks for connection and guidance from God, believes that God forgives, and carries their religious beliefs into other areas of their life, the more likely they are to forgive.

That forgiveness also changes our brains. In a study at the University of Pisa, participants were recruited who scored high on the ability to imagine scenarios vividly. While in the fMRI scanner, they were given a series of scenarios in which an imaginary person was wronged, and they were asked to either imagine forgiving the offender or lean into resentment and imagine revenge. All participants reported being able to feel anger and frustration while imagining the hurtful scenarios, and they reported that the more they felt able to forgive, the more they experienced a sense of relief. Moreover, forgiveness activated their brains, with greater signal seen in the *dorsolateral prefrontal cortex*, an area associated with control over emotion and behavior, the *precuneus*, which processes what we perceive is relevant to our own lives, and the *inferior parietal lobule*, which represents what is going on around us. Moreover, a general tendency to forgive is associated with lower average activation in the dorsomedial prefrontal cortex as well as the precuneus and inferior parietal lobule. Along with changing our brains, forgiveness reduces our anxiety and depression while increasing self-esteem and hope. It frees our brains from rumination and negative thoughts.

My father's choice to forgive, as difficult as it was, likely improved not only his ability to continue working at a Christian college but also his health and well-being. Although prospective clinical trials of forgiveness on health have yet to be done, over one hundred published studies show that forgiveness is associated with better physical health, including by reducing cortisol levels, lowering blood pressure, and improving cardiovascular health. As the common saying goes, "Resentment is like swallowing poison and expecting the other person to die." *Our beliefs, to the extent they help us choose forgiveness and let go of resentment, can change our brain activity and help us live happier and healthier lives.* Freeing ourselves from negative thoughts and emotions like resentment allows us to more fully experience the positive—including gratitude, resilience, and hope—along with sometimes surprising benefits to our brains and health.

Gratitude

My colleague Joel Wong, a professor and former department chair of Counseling and Educational Psychology at Indiana University, and I have two things in common. Our then-five-year-old daughters both learned to play violin in our university's string academy for young children, which is where we first met, and we both have an interest in gratitude research. Much has been written about the benefits of gratitude in recent years, and Wong is a leading expert. Because gratitude correlates with religiosity, by studying gratitude we can better understand the social and neural mechanisms through which spiritual beliefs affect well-being.

In 2012, Joel approached me with a proposal to study gratitude together—he would look at clinical effects of gratitude on mental health, and I would examine how gratitude might change the brain. Soon we were funded by the Greater Good Science Center at the University of California, Berkeley. We recruited three hundred participants, all of whom were receiving psychotherapy for mental health concerns. We divided the subjects into three groups. One group wrote letters of gratitude, though they didn't need to give the letters to addressees. The second group wrote about their most stressful life experiences. The third group didn't write any letters. There was no immediate difference among the groups in their mental health scores, but one to three months later, the group that wrote gratitude letters had significantly better mental health scores than those in the other two groups. The simple act of writing letters of gratitude had helped their mental health in a lasting way. Our next step was to examine how gratitude might be affecting subjects' brains.

Three months after the letter-writing intervention, we scanned a subset of the gratitude letter writers and control participants (those who didn't write letters) while they did what we called a "pay it forward" task in the fMRI scanner. During scanning, the participants were told that someone had just given them a certain amount of money and didn't want it back. Instead, if they felt grateful, they could choose to give some or all of the money to another person in need or to a local charity of their choice, such as a food pantry. We promised to (and did) give real money to the charities chosen. Our participants on average gave over 50 percent of the money they were given to a charitable cause. We asked them how grateful they felt, and also how much they simply desired to help the charity or would feel guilty for not doing so. We found that how much gratitude the subjects felt correlated with brain activation in the *parietal cortex*, which helps us process our surroundings. The gratitude was changing their brain activity.

It turns out that gratitude is also a powerful engine of neuroplasticity. The gratitude letter writers had more sensitivity to gratitude in their ventral medial prefrontal cortex than the controls did, even three months after they did the gratitude letter-writing exercises. For those who had written the gratitude letters, the more gratitude they reported during the scanner task three months later, the more that gratitude activated their brains. The control participants didn't show as much sensitivity. This was a remarkable finding—*a simple act of gratitude caused lasting changes in the brain even three months later.* The neuroplasticity effect had worked: Their brains had changed how they responded to gratitude, and their sense of well-being increased.

This finding that gratitude changes the ventral medial prefrontal cortex is noteworthy because it is the same brain region that is more active in people who are more optimistic. It activates when people imagine positive future events or recall positive past events, and its activity can also tamp down amygdala activity related to negative emotions. The more optimistic people are, the more active this region becomes when anticipating positive future events. The sensitivity to good entailed by gratitude practices and associated optimism can set us up to stay resilient and weather hard times better.

Grit

After visiting with James and the local prayer group and experiencing so much peace shortly after my diagnosis, I began to travel elsewhere in the US, visiting other groups that offered to pray for my healing. These experiences of receiving prayer left me with a greater sense of well-being and degree of hope. By November 2003, I had developed a guarded optimism that the prayers might be helping to heal the brain tumor, and I carried this into the three-month follow-up MRI.

The day after the MRI, in a small courtyard outside Barnes-Jewish Hospital in St. Louis on a sunny fall day, I again opened a bulky manila envelope containing my latest scans and readings. Were the prayers having any effect on the tumor? Was it shrinking, or even totally gone? My eyes scanned the reading and landed on the result: "Mass in left uncus involving hippocampus head and amygdala with hyperintense FLAIR signal without enhancement. Findings most consistent with a low-grade astrocytoma, ganglioglioma, or oligodendroglioma." Which is to say—the brain tumor was still there. In that moment of disappointment, I had a decision to make: Do I give up on pursuing healing and prepare to die? Or do I double down and continue to pursue healing prayer and whatever other help I could find, despite the cost in both travel expenses and precious time with family? I decided then that I would either find healing or die trying.

In retrospect, making a decision that pointed me so firmly to the future, even if the future was unknowable, might have been the exact right decision to make even for non-supernatural reasons. *Grit* is a personality trait defined by having both a strong sense of purpose and persevering toward it despite difficulty. When things go wrong, instead of giving up, we are resilient. We figure out how to overcome challenges and still move forward toward achieving our goals. Grit has important health benefits. And it turns out that our beliefs help give us grit. Those with a higher degree of religious or spiritual beliefs on average have higher levels of resilience. Our beliefs can provide a sense of meaning as well as hope that can motivate us to keep going.

Research has shown that grit is associated with better health outcomes. In a study of four hundred college students with chronic medical conditions, higher levels of grit were associated with better quality of life. Students identified as having grit set goals and pursued them despite setbacks. They followed up with medical treatments more consistently, keeping all their appointments, and followed doctors' prescriptions, whether for medicine or diet change, all of which aided their well-being. And though there was no significant evidence of a direct link between this kind of activity and better physical health, nevertheless, researchers did observe that those with higher levels of grit experienced better health, both physically and mentally.

How, then, does grit improve health? There are two main ways. First, grit means we approach life with goals. We have a sense of purpose, and part of that purpose is to take care of ourselves and pursue well-being. Having a purpose and pursuing it with clarity leads to better emotional well-being. Second, instead of reacting negatively to problems, individuals with a grit mindset switch to problem-solving. Our drive is to push beyond the problems and seek solutions. Notably, satisfaction comes from engaging in the process, not just in achieving the desired outcome—hence the aphorism about life being a journey and not a destination. Specifically, we can find a sense of emotional well-being in reflecting on the progress we're making toward our goals.

Neuroscience research has shown that our grit, or resilience, makes a positive difference in our brain function. One way to study the difference that grit makes to both our emotional and physical well-being is by looking at mice subjected to mild chronic stress. Unpredictable chronic mild stress in mice can produce depression-like symptoms. In one study, researchers subjected a group of mice to social stress. They put a mouse in a cage with a more socially dominant, aggressor peer mouse, resulting in an encounter in which the less dominant mouse was either attacked or displayed a submissive posture. About half the mice who experienced social defeat displayed resilience—they kept interacting socially despite their harrowing experience with their dominant peer. The other half of the socially defeated mice, however, primarily opted out of subsequent social interaction. They simply didn't have the same kind of resilience as their counterparts and avoided their peers as a result of the negative experience with the aggressor mouse. Their social withdrawal, a direct result of social stress, left them depressed.

The resilient mice didn't just benefit in terms of a better social life. They physically benefited from their social grit. They had lower inflammation and a lower cortisol increase than their less resilient peers. The mice's experience with social stress also left the resilient mice with less severe learning and performance effects than it did the less resilient mice. The resilient mice were better able to learn new information despite the social stress they'd experienced, instead of the weaker learning-related changes seen in the less resilient mice. That's because the resilient mice's hippocampi were less affected by the stress.

It's not just mice. Ongoing stressful events have a harmful effect on humans as well. And it's not only our hippocampi that are affected. Stress also affects our frontal lobes, which are essential for things like pursuing goals, planning, and regulating our emotions. It's why when we feel continuous stress, we may find it hard to calm down, think clearly, and plan our next steps. Over time, stress can cause some connections among neurons in the frontal lobe to retract, making them less able to communicate with other neurons. We think less clearly when we are repeatedly stressed.

Stress and lack of resilience are a vicious cycle. When we experience a stressful event, the stress causes a cascade of neural changes. Our cortisol levels increase. It becomes easier to learn fearful associations rather than positive associations. We withdraw from things. Also, the cortisol and stress more generally rewires our frontal cortex, which makes it harder for us to stay focused and pursue more meaningful goals.

In one experiment, healthy subjects and those with major depressive disorder were put in a stressful situation, like having to do mental arithmetic while their hand was submerged in ice water. Now, I know for some of us, math might be a piece of cake, but for many people, having to do mental arithmetic, especially when our hand is freezing, is a stressor. The already stressed subjects were then asked to play a gambling game in which they would periodically win or lose a small amount of money while their brain activity was measured with fMRI.

For most people, losing a gamble is yet another stress point, even when playing for peanuts. We want to win, and our brains want us to win. When something bad happens, like losing a bet, our amygdala usually reacts. Normally the frontal lobe dampens the amygdala response. With stress, however, the frontal lobes don't exert as much control over the amygdala. Stress removes the brakes from the amygdala, metaphorically speaking, so that it responds more strongly when bad things happen. So, when participants lost their bets, on top of freezing hands and brains that were already stressed from doing mental arithmetic, their frontal lobes weren't dampening the amygdala response as much as they would otherwise. Apparently the connection from the frontal lobes to the amygdala was somehow broken. The healthy participants' brains began to look a lot like those of the participants with major depressive disorder: Their responses weren't being braked and their emotions were less well regulated.

The good news is that broken neural connections may be able to re-form after the stress is resolved, restoring our ability to pursue our plans effectively and to emotionally regulate ourselves. Once the stress lessens, healthy individuals, especially those who have developed resilience or grit, are likely to show brain activity patterns that revert to normal.

The solution is to break the cycle. Grit is not just a static personality trait. *We can develop grit by intentionally setting goals, reflecting on our own and others' experiences of grit, and adopting a growth mindset.* A *growth mindset* is to believe we can develop our talents through hard work, strategies, and input from others, as opposed to a *fixed mindset*, where we believe our talents and abilities are innate gifts that cannot be meaningfully improved. Studies have shown that people with a growth mindset achieve more than people with a fixed mindset. In the adoption of a growth mindset, once we've set our goals, we take whatever steps we can to pursue them. We then look for satisfaction not just in achieving goals but in making progress along the way. Setting a goal takes us out of the space of simply reacting to difficult events in life with fear and stress. By being proactive instead of reactive, we can increase resilience, help our brain function better, and ultimately develop a deeper sense of well-being.

While the evidence suggests that grit can improve our health, I offer these findings tempered with a degree of humility, recognizing that not every battle can be won. Why some find healing and others don't ultimately remains a mystery, and no one can be faulted for suffering or losing a health battle. I was fortunate to find friends who both encouraged me not to give up hope and who stood by me with compassion and unwavering support when my medical results did not show what I had hoped. I strive to offer the same support, while encouraging the same hope, for those in my life who need it.

Hope

When I was six years old, Aunt Katie, my dad's only and much-loved sister, was diagnosed with breast cancer. Aunt Katie was in her early thirties, recently married, and looking forward to having a family of her own, when suddenly, one otherwise unremarkable day, she found a lump in her breast. We were at home when the phone rang. Dad answered.

"What's going on?" I could tell he was concerned.

"It's already spread to the lymph nodes and beyond. They're recommending lumpectomy. After the surgery, they want to follow up with chemo, and then radiation. And, Billy, even if they do all that, they can't assure me I'll be cured."

"What do you want to do?" he asked.

"I'm going to get the surgery. And the chemo and radiation. I'm going to do it all. I'm not giving up."

After the call, my father sank down into our olive-green

1970s easy chair. We gathered around him concerned as he bowed his head, radiating shock and sorrow. In low, measured tones, he uttered what he would go on to repeat often over the coming months: "Heavenly Father," he prayed, "we know you can heal Katie and relieve her pain, and we ask you to do so in Jesus's name. But not our will, but yours be done. Amen." It was a prayer filled with both hope and resignation. The prayer would be answered, even if the answer was no.

For her part, Katie appreciated the support.

"I felt your prayers, Billy. I still do. They mean a lot."

I never got the sense from my dad that he believed God would perform a miracle in response to his prayers. But we hoped with every fiber of our being that we could help make Aunt Katie better, even when deep down we knew there was little we could do. Our prayers were a way in which we might contribute. Aunt Katie reiterated that she felt our love and prayers and concern, and that we helped her not to give up hope.

Aggressive treatments took a harsh toll on Aunt Katie physically and emotionally. Then, much to our surprise, and despite the doctors' grim prognosis, she experienced a full recovery. And her breast cancer never returned.

When I was diagnosed, Aunt Katie's story wasn't at the forefront of my mind, though it hovered in the background. I'd always wondered whether my dad's prayers contributed to her recovery and, if so, how. Now I wonder whether knowing she healed influenced my own recovery. Perhaps the knowledge of her survival, buried deep in my cognitive map, guided my own faith that recovery was possible despite having received a terminal diagnosis of my own.

The memory of Aunt Katie's healing gave me hope, but some memories may not be so helpful. I have another, very specific memory of a senior pastor visiting Candy and me at home shortly after I was diagnosed and telling me a story of a woman who was dying of a brain tumor. He had the best intentions—to spare us from disappointment and despair. But hearing about someone else's death from the same disease with which I'd just been diagnosed was unnerving. His story didn't deeply traumatize me, but it was an emotional event. It affected me to the point that I remember his words to this day, and it readily comes to mind if I'm recalling the people I know who have died of cancer. When it's an emotionally heavy memory, there is more activity between the amygdala and medial temporal lobe memory regions. That increased activity is the attention we're paying to the memory.

When we predict our future, we draw on our past. *But which past will we choose*? We have many memories—of good times, of bad times, of successes and failures. Memory recall depends on what we pay attention to and is controlled by our ventrolateral prefrontal cortex. The memories our ventrolateral prefrontal cortex recalls in turn affect our predictions according to the availability heuristic discussed in chapter 3. And as we saw in chapter 4, healing prayer focuses our attention on our concerns and helps us find useful information faster.

After becoming involved with healing prayer groups, the stories I heard and the experiences I witnessed changed completely. My attention was no longer hyperfocused on terminal cancer stories. Now, stories of health and recovery were the context within which I operated. As I focused on healing and solving the problem of my survival and recovery, stories of healing gained priority, many of which I pulled from memory. It was at this later stage that the story of Aunt Katie's recovery came back to me and helped orient my new narrative around the possibility of healing. This more positive memory, not the memories I had of terminal brain tumor patients, now guided my future predictions. That in turn gave me renewed hope.

Clinical studies have also shown how prayer can increase our capacity for hope. Chronic kidney disease patients require regular dialysis, which can make life difficult with little or no prospects of improvement. In a recent study, researchers divided dialysis patients into two groups. One group listened to six-minute recorded prayers during dialysis sessions every other week, over a total of three dialysis sessions. Relative to the control group, those experiencing the prayers later showed significantly greater optimism, made more plans for their daily life, and expressed more sentiments that their life was useful and valuable. In other words, *the prayers helped by giving them more hope.*

Hope also matters to our well-being and ability to thrive. In 1972, Professor of Psychology Martin Seligman at the University of Pennsylvania, who is credited as the father of positive psychology, carried out a series of experiments in which animals lost hope. When dogs were subjected to an electric shock from which they could not escape, they initially struggled, but after a while, they gave up. Even when it later became possible to escape, they no longer tried, and they suffered further shocks that they could have easily avoided. The dogs in Seligman's experiment suffered what is called *learned helplessness.*

This learned helplessness, as it turns out, can be fatal. Rats that are thrown in water and forced to swim can survive sixty hours. But if a rat is first held in the hand and squeezed so that it cannot escape and finally stops struggling, it has now learned to be helpless. If a helpless rat is dropped in water, it will drown within thirty minutes—a reduction in survival time of 99 percent.

Learned helplessness can affect us as well. In a follow-up study, Seligman and his research colleague Donald Hiroto, a clinical psychologist, gave college students unpleasant experiences from which they could not escape, such as loud noises or cognitive tasks that were impossible to solve. Later, those who had experienced inescapable loud noises and unsolvable problems were given unpleasant situations they could escape from, and problems they could solve. They performed much worse, because they had learned that trying was futile. The fact that we are able to overcome a challenge is not enough—we need hope in order to turn our potential into reality.

Looking at the bigger picture, our beliefs, including religious and spiritual beliefs, can help us not only to overcome mental health concerns but also to experience an overall sense of well-being. Social support, forgiveness, gratitude, grit, and hope all help us face challenges with optimism and persistence, increasing the likelihood of better outcomes. Moreover, as we'll see in the next chapter, our beliefs can also have positive effects on our physical health.

6

How Beliefs Affect Physical Health

ANITA MOORJANI IS A *NEW YORK TIMES* BESTSELLING AUTHOR and cancer survivor. In her book *Dying to Be Me*, she describes being diagnosed with Hodgkin's lymphoma and battling it for four years. While in the hospital with organ failure and a subsequent coma in 2006, she had a near-death experience that transformed her understanding of both her cancer and her life. She describes an out-of-body encounter during which she felt "pure, unconditional love . . . the deepest kind of caring, and I'd never experienced it before." After she regained consciousness, she recovered. The cancer disappeared in weeks. Moorjani's near-death experience is not uncommon, but what accounts for the sudden, seemingly miraculous recovery? Or more specifically, could changes in her beliefs through her near-death experience have helped her fight the cancer?

Moorjani's profound spiritual experience seems at the center of her remarkable recovery. There is notable and extensive evidence that people who are more religious tend to enjoy better health. Harold Koenig, the medical doctor and professor at Duke University who conducted the prayer study with Peter Boelens whom we met in chapter 4, is a leading expert on how religion and spirituality affect our health. In a landmark 2012 review, Koenig showed that across hundreds of studies, greater spirituality and religious adherence overwhelmingly correlates with better health for a range of conditions, from anxiety and depression to cancer and heart disease. Koenig advocates for taking a patient's spiritual and religious history into account as part of standard medical care—not to push particular beliefs but to improve medical outcomes by working with patients and their beliefs to address their spiritual needs. As we gain understanding of the complex ways our beliefs influence health conditions—from cancer to pain—we, too, may enjoy better physical health.

How Beliefs Affect Cancer

Bruno Klopfer, a clinical professor of psychology at the University of California, Los Angeles, gave a presidential address in 1957 at the Society of Projective Techniques annual meeting in New York City, in which he related a story of a cancer patient with advanced lymphosarcoma. The patient, whom Klopfer called Mr. Wright, was in palliative care; he had tried and failed all therapies available at the time, with large tumors the size of tennis balls visible throughout his body. Wright heard that a new experimental cancer treatment called Krebiozen—

a combination of the amino acid creatine and mineral oil—was to be tested at the hospital where he was being treated. Wright begged his doctor to be allowed to try the treatment, even though by his doctor's estimate, Wright had two weeks left to live at best. Nevertheless, the doctor agreed.

The first injection was administered on a Friday, and by the following Monday, even before the second injection was to be given, Wright was out of bed and happily walking around. The doctor observed that "the tumor masses had melted like snowballs on a hot stove . . . half their original size!" None of the other patients being treated with the same experimental drug showed any improvement. Wright continued to receive the injections three times per week, and after ten days, he was discharged in seemingly good health having experienced a remarkable recovery. As it turned out, within a few months, Krebiozen was shown in a larger study to be useless, a verdict that stands to this day. When Wright learned that the drug was found to be ineffective, his ebullience gave way to despair, and the tumors quickly returned. As the doctor describes, he then "saw the opportunity to double-check the drug and maybe too, find out how the quacks can accomplish the results that they claim . . . I deliberately took advantage of him [the patient] . . . for purely scientific reasons."

The doctor informed Wright that "a new super-refined, double strength product is due to arrive tomorrow." Wright took greatly renewed hope from this, and a few days later, the doctor injected the patient with pure fresh sterile water, with no active ingredients except the excitement and fanfare accompanying the injection of this fresh batch of supposed wonder drug. The doctor described that "recovery from his second near-terminal state was even more dramatic than the first. Tumor masses melted, chest fluid vanished, he became ambulatory, and even went back to flying [his personal aircraft] again. After another symptom-free two months, the American Medical Association itself announced that Krebiozen was worthless as a cancer treatment. Upon hearing this, Wright's faith in the drug evaporated, and he was readmitted to the hospital in extremely poor condition, where he died two days later.

The ethics of medical treatment, like cancer treatments themselves, have come a long way in the last sixty years. Although the results of the Krebiozen case study offer insight, such an experiment would never be done today in the way that it was then. By modern standards, it would be unethical to provide experimental medical treatment without the patient's full informed consent or merely to indulge the doctor's hubris in proving "the quacks" wrong. In the United States, ethically dubious research was conducted routinely until the 1970s. The Tuskegee Syphilis Study, in which researchers deliberately infected poor African American men with syphilis without their knowledge, consent, or subsequent treatment, awakened the conscience of the nation in 1972 when details of the study became public. By 1978, the National Research Act empowered a committee of experts to produce the *Belmont Report*. This report formalized a code of ethics in medical research that all subsequent medical research would have to adhere to under federal regulations. In particular, medical research could only be performed with patients who provided fully informed consent, meaning they understood the research procedures; they understood the potential risks and benefits to themselves; and they agreed to participate freely and without coercion. Moreover, research could only be performed where the potential benefits outweighed the potential harms. These ethical principles today govern all aspects of medical research, including research on cancer.

While such research continues to yield new and better cancer treatments, the human body has a remarkable ability to fight cancer on its own, and that ability depends on the immune system specifically. For example, people who have organ transplants have to take immunosuppressive drugs to prevent the body from attacking the transplant; immunosuppression increases risk of cancer. Some of the earliest cancer treatment methods involved primitive attempts to activate the immune system, and while as with the Krebiozen case they wouldn't be performed today, it's notable that some of them might actually have been effective. In the late 1700s, physicians deliberately infected the leg of a breast cancer patient with pus in order to test her immune system response. Their hope was that the infection would stimulate her immune system to fight her breast cancer. The results were startling: The experiment worked! Or at least in a fashion it did. What started out as a mild infection worsened. Yet the worsening infection apparently did stimulate her immune system as her breast cancer soon went into remission. No one now would recommend deliberately infecting a patient with unknown germs, of course, but it was a fascinating result, and one that led to subsequent attempts to treat cancer by infecting patients with specific strains of bacteria and viruses. In particular, in 1891, researcher William Coley administered inactivated bacterial strains named Coley's toxins to cancer patients and also saw a degree of success in treating cancer.

More sophisticated attempts at activating the immune system with the goal of destroying cancer continue. In 1988, Steven Rosenberg, Chief of Surgery at the National Cancer Institute in Bethesda, created specially modified immune cells called tumor-infiltrating lymphocytes, which were able to attack melanoma cancer cells better than a patient's own immune cells. The patients they treated this way saw their cancers regress, in some cases for over a year. Over the last decade, there has been tremendous progress in developing immunotherapy against cancer. This approach uses and strengthens the body's own immune system to fight cancer, as described in the book *The Breakthrough*, by Charles Graeber. In it, Graeber explores one of the most exciting developments in cancer treatment: checkpoint inhibitors. In simple terms: The immune system doesn't always attack cancer cells because they have molecules called checkpoints, which are like official identification that the cell is a normal part of the body and shouldn't be attacked. Much like airline personnel have special identification that allows them to go onto the tarmac without getting arrested, cancer cells display their checkpoint molecules to shut down attacking immune cells. The new checkpoint inhibitor drugs shut down the checkpoints, confiscating their official identification so that the immune system can quickly detain them. Other approaches using chimeric antigen receptor therapies (CAR-T) create specially modified immune cells that simply ignore checkpoints and attack cancer cells despite their official identification. These checkpoint-ignoring immune cells can be created easily in a lab now. Cancers like melanoma that were previously incurable can now be treated.

All this gives us some hope that by strengthening our immune systems, we might help our bodies fight off tumors like the one I had. But this begs the question: Why can't the immune system always fight off cancer? It is, after all, the immune system's job to detect pathogens such as cancer cells and protect us from disease. There are three main reasons. First, cancer cells are opportunistic. If the immune system is already weak, for example, in people with organ transplants who are immunosuppressed, cancer often grows more easily. Second, cancer cells are crafty. They hide from the immune system, camouflaging themselves as if they were normal, healthy cells, and then use their undercover status to continue their attack. Third, cancer cells are powerful. Once there is a critical mass of cancer cells, the cancer itself may weaken the immune system. As the number of cancer cells increases, even if they have not yet metastasized, they prevent immune cells from growing or functioning properly, so they never have a chance to destroy the cancer.

As we saw in chapter 2, our immune system can be strengthened or weakened by fluctuations in our cortisol levels, and our cortisol levels in turn are influenced by our brain. Kirk Warren Brown, a professor of psychology at Carnegie-Mellon University, and colleagues published a study of cancer patients in 2003 that looked at the role of mental well-being in how long people survive cancer. Brown followed patients for ten years. The study looked at a range of factors, such as how healthy patients felt, how happy, how anxious, how depressed, and how in-control they felt and whether those factors could be linked to a longer survival time. Some of the factors had ambiguous results, but what was unambiguous was the role of depression: The fewer symptoms of depression people had, the longer they lived, regardless of other factors. Even patients with a worse kind of cancer lived longer if they weren't depressed.

It feels intuitively true that the people we know who face terminal illnesses with optimism and hope, rather than collapsing into depression, live longer. A 1997 study noted that depression is associated with excessive activity in the *hypothalamic-pituitary-adrenal (HPA) axis*, as discussed in chapter 2. This disproportionate activity causes an overabundance of cortisol to circulate in the body. Cortisol in turn weakens the immune system, preventing it from producing immune cells and chemicals that would otherwise attack invaders. Depression, or really anything that's going on in our brain, directly influences our hormone system and thus our ability to fight cancer.

In fact, there is a whole field of scientific research called psychoneuroendocrinology, which studies how the brain influences hormones, which are part of the body's endocrine system. Stress in rats increases stress hormones like cortisol, which then increase tumor growth, especially in older rats. So, at least in rats, stress worsens cancer. It's not too far a leap to see why the stress associated with depression might have the same result in humans. In a population of cancer patients, those who are stressed or depressed have increased cortisol levels relative to those with less stress and depression, and this is associated with shorter survival times. A growing body of research shows a high correlation between elevated cortisol levels and the progression of cancer. *Our spiritual beliefs and practices, to the extent they reduce our stress, may help us avoid or even survive cancer.* While cancer is life-threatening and the second-leading cause of death after heart disease in the United States, chronic pain also lessens quality of life for many Americans. In 2023, 24.3 percent of US adults reported chronic pain.

Pain

It was May 2024, on a pleasant, sunny day in downtown Baltimore. An African American woman in her sixties, whom I'll call Nora, visited the University of Maryland Family and Community Medicine Clinic for her regular medical care. As she walked through the glass door of the redbrick building and sat in the sparsely decorated waiting room, a friendly woman approached her and asked, "Would you like to be part of a research study on pain and anxiety?" Nora had been suffering for the last seven years from severe pain in her legs, arms, and chest, which she rated as eight out of ten on a pain scale. Any prospect of relief sounded appealing. Besides, the researchers were offering modest compensation for being part of the study. If Nora agreed, someone would randomly assign her either to listen to relaxing music or be prayed for by a Christian volunteer for five minutes after her medical appointment. The researchers would ask her some questions over the next two months about her experience; she could stop participating at any time if she felt uncomfortable. Nora agreed and signed the informed consent form under the clinic's bright fluorescent lights. Although she was not affiliated with any particular religious tradition, Nora felt that she might as well try prayer. The researchers, for their part, included people in the study regardless of their religious beliefs.

Nora's medical appointment ended, and a researcher took her to another exam room in the clinic. A volunteer began praying with Nora for five minutes while Nora sat quietly. The volunteer offered simple prayers in a gentle voice, asking God to heal the pain and softly commanding the pain to leave, invoking the authority of Jesus. By the end of the five minutes, Nora wrote, "It feels like a miracle or something happened. I still have pained [*sic*] but it is now lessened . . . It worked! This is the 1st time I've experienced that . . . The pain seems to be dropping—radiates down legs. Cannot stand or sit for a long time. It lifts my spirits + gives me hope."

Although Nora's experience does not prove whether God did or did not intervene on her behalf, the fact is that she experienced a meaningful relief of pain as she received prayer. In the next chapter, we explore how placebo and other mind-body effects might explain Nora's experience. But for Nora and many others in the study, the effect was very real, powerful, and lasting.

Nora was one of 183 participants in this research project that ran from May through August 2024. It started in 2018 as a conversation I had with a friend of mine, Dr. Kate Jacobson, a professor at the University of Maryland School of Medicine, Dr. Jacobson is a family practice physician, a teacher, and a scholar with a passion for patient care and research on evidence-based practices. We were discussing the relative absence of research on the effects of prayer for healing and what we could do about it. The idea took shape over time; when the COVID-19 pandemic began, we hit pause. But once the pandemic shutdowns eased and we were able to start new projects, we gathered a team of mostly volunteers and designed a study. Our questions were straightforward: Could brief, in-person healing prayer improve physical pain symptoms as well as symptoms of anxiety? And if so, were the improvements temporary or lasting?

When we analyzed the data from all 183 participants, we found that those who listened to soft music for five minutes and those who received prayer for five minutes both reported reduced pain afterward. Those who had received prayer, however, showed a more dramatic reduction in pain, about two points lower on average on a ten-point pain scale, compared with those who had listened to music. More remarkably, when we followed up with subjects two weeks later, those who had received prayer still had significantly lower pain scores than those who had listened to music. *Not only had the healing prayer helped them in the moment, but the effects lasted at least two weeks*. Even more interesting was the finding that it didn't seem to matter whether the patients who received prayer had the same Christian faith as the volunteer who prayed for them or not. The level of pain reduction the patients experienced after prayer was the same. It also made no difference how much patients had attended religious services, their age, household income, or gender, nor even whether and how much they expected that the healing prayer might help them. The prayer experience seemed to help everyone.

We also found that patients with anxiety benefited, showing lower anxiety scores immediately following prayer. Anxiety scores from patients who had received prayer were also lower compared with those who listened to music at the two-week and six-week marks—a lasting effect.

We also considered whether the in-person nature of the prayer experience affected results. We recruited over eighty additional participants to receive prayer remotely (or listen to relaxing music, for the control participants), via a video call. They knew they were being prayed for and what the contents of the prayers were, as they could see and hear the volunteer praying for them over the video call, but they were not physically present with the prayer volunteer. These "virtual" prayers had no significant effect on the ten-point pain or anxiety scores relative to the control condition. In addition, we found that the pain reduction from in-person prayer was significantly larger on the ten-point scale than effects on pain reduction from prayer via a video call. Nevertheless, that wasn't the whole story. The patients' written comments more broadly showed they felt significantly better in terms of both alleviated symptoms and greater well-being after prayer than the control condition. This was true both for those who received prayer in person and those who participated by video call. In other words, *prayer for healing seems to help reduce pain and anxiety symptoms, and improve well-being in all participants, especially for those who received prayer in person*. In a COVID-era world, much interpersonal interaction has moved online, but there still seems to be no substitute for the health benefits of in-person connections. Sensing something similar, medical professionals have recently renewed calls for physical touch as an indispensable part of the doctor-patient relationship.

Prayer Studies

Our findings follow a handful of experiments over the years that have looked at prayer and its effects on health. These studies can roughly be divided into two categories: *distant intercessory prayer* and what Candy Gunther Brown calls *proximal intercessory prayer*. Distant intercessory prayer refers to prayer for healing from a distance, in which the person receiving prayer is unaware of the fact that they are being prayed for. Proximal intercessory prayer refers to face-to-face prayer for healing, with one person offering prayers for healing of the other person. The idea of studying healing prayer effects from a scientific perspective goes back to the late nineteenth century, when skeptical physicist John Tyndall proposed a prayer-gauge test, in which people in one ward of a hospital would be prayed for, while those in another ward would not. The idea provoked controversy and was never pursued.

In 1988, Randolph Byrd, a cardiologist at the University of California, San Francisco, conducted a study of distant intercessory prayer for coronary care patients. Byrd found that those who received prayer (without knowing they were being prayed for) suffered fewer complications, but the study was criticized as depending on a way of categorizing the severity of complications that had not been validated and might have been determined after the fact, casting doubt on the conclusions.

In 2006, Herbert Benson carried out a well-funded attempt to replicate Byrd's findings in coronary care patients, but drew the opposite conclusion: Those who were prayed for without knowing it fared no better than those who were not prayed for. Moreover, those who were told they were being prayed for actually fared worse, meaning they had more complications than those who were prayed for but not told. These findings were also criticized on methodological grounds, because the volunteers who provided the prayers did not themselves believe that prayer would improve symptoms. This in turn raised the question of whether the study actually investigated the effect of prayer as it is commonly practiced, since individuals who pray for healing typically believe that it may be effective. A Cochrane review, the gold-standard of systematic research, concluded in 2009 that the evidence for healing prayer was mixed at best and discouraged further studies. Despite the Cochrane review, my thoughts on the matter were still up in the air. Regardless of mixed results and being discouraged from further study, I felt compelled to do just that.

Perhaps the most controversial aspect of the distant prayer studies is that if positive effects were found, it would be difficult to explain apart from supernatural or quasi-scientific, nonlocal mechanisms. After all, the patients had no way of knowing they were being prayed for, so mind-body interactions would not be viable explanations for any effects. Those who believe in absolute naturalism would have no reason to expect such effects nor a clear way to account for them. This leads us back to the distinction between distant versus proximal (or remote versus in-person) intercessory prayer. Studies of proximal intercessory prayer need not offend the sensibilities of absolute naturalists in the way that distant prayer might. Proximal prayer by its nature makes clear to recipients that they are being prayed for. In clinical trials of pharmaceuticals, typically both the doctors and the patients are "blinded," meaning neither one knows whether a particular patient gets the active drug or an inert sugar pill. With in-person prayer, blinding is lost. Both the person praying and the person receiving prayer know the prayer is happening. That means any effects of prayer that may be found can potentially be explained as mind-body interactions, perhaps as prayer reduces feelings of stress, as seen in chapter 2, or possibly via placebo effects we explore in the next chapter. Proximal intercessory prayer can be studied as a way of learning about possible psychological or mind-body mechanisms, as well as to learn about how spiritual beliefs may affect health.

Prior to the study my colleagues and I did in Baltimore, there had been a handful of other studies of in-person healing prayer. In the 1970s, celebrity Catholic priest Father Francis MacNutt traveled the world praying for healing before crowds of fifty thousand people, and his 1974 book *Healing* became a bestseller. By the time I met MacNutt in 2006, decades of running had worn down his knees and made walking painful, but he continued to pray with people for healing until his passing in 2020 at age ninety-four. As a Harvard graduate who studied medicine, MacNutt had long understood the importance of scientific investigation. In 2000, he teamed up with Dr. Dale Matthews to study the effects of prayer on patients with rheumatoid arthritis. Half the patients received a three-day in-person healing-prayer experience that consisted of six hours of group instruction in the theology of healing prayer and six hours of receiving prayer individually. Matthews found that six months later, those who had received prayer had fewer tender joints and better daily functioning, although curiously, their biological markers of inflammation showed no difference between groups. Half the patients were randomized to receive distant intercessory prayer, such that the patients were blinded to whether they were the subjects of prayers. Prayer from a distance had no effect on any of the measured outcomes.

How Beliefs Change the Brain to Relieve Pain

Religious and spiritual beliefs are often expressed through practices such as prayer and meditation, and these practices can alter our brain activity to reduce pain, sometimes in dramatic ways. John Foxe's classic *Foxe's Book of Martyrs*, first published in 1563, records many instances of individuals being tortured and killed for their religion while showing no signs of pain. When I visited my cousins in England twelve years ago, they told me a story of a local minister from long ago who had been martyred for his Protestant faith just up the hill from their house. In the year 1555, Rowland Taylor was killed by local authorities for refusing to recant his religion. As he climbed into a barrel of flammable oil and stood next to the stake where he would be burned to death, he reportedly folded his hands, looked up, and prayed, staying motionless and silent as the fire was lit and burned him. He seemed impervious to the pain.

Ordinarily, being burned alive would result in uncontrollable screams of agony. How could the unfortunate minister have seemingly experienced little or no pain?

In 2008, Katja Wiech, Irene Tracey, and colleagues at Oxford University tested how religious beliefs might provide pain relief. To do this, they scanned a group of devout Catholics and a group of atheists with fMRI, while giving them mild electrical shocks to the back of the hand. About thirty seconds before the electric shock, participants were shown a picture of the Virgin Mary or a control picture of a woman with similar features but not identifiable as the Virgin. The atheists rated the pain as the same after seeing each picture, but the Catholic participants rated the pain as less intense after seeing a picture of the Virgin Mary relative to seeing a picture of an unknown woman. The Catholics rated the picture of the Virgin Mary as religiously significant, saying they had been affected by it. The more they felt affected by seeing the religious image, the more active was their midbrain, a small region at the base of the brain. The atheist participants felt no such effect. The brains of the Catholic participants showed more activation, especially in the ventral lateral part of the prefrontal cortex, or VLPFC, when they saw the religious image, and the degree of activation correlated with the degree of pain reduction. The Catholic participants reported that the religious image helped them feel "calmed down and peaceful," "taken care of," and "safe," which might have helped them experience the pain as less emotionally bothersome. The VLPFC activation might have helped the Catholic participants suppress the pain signals or perhaps experience them as less disturbing. How the VLPFC and related brain areas can suppress pain is a question we return to in the next chapter, "Placebo Effects."

In part 2, we have seen that our beliefs, including spiritual beliefs and practices as well as our beliefs about the world more broadly, can have profound and surprising effects on our health and well-being. These beliefs and practices provide perspectives on the world, reframing and relieving stress, and fostering social connection, forgiveness, gratitude, resilience, and hope. Spiritual practices such as prayer can help us overcome anxiety, depression, and addiction. The social experiences and support facilitated by our beliefs can also change our brains in lasting ways: lowering cortisol, increasing oxytocin, and improving both our mental and physical health, especially when it comes to battling cancer and overcoming pain. We can trace much of how this works through mechanisms in the brain. In part 3, we dig deeper and find that our beliefs can unlock even more dramatic healing experiences, some of which can be explained by neuroscience, and some of which remain a mystery.

PART III

More Healing Is Possible

7

Placebo Effects

When I was ten years old, my father, in the middle of a karate session in a dilapidated gymnasium, dressed in his white gi and sporting a blue belt that reflected his intermediate level of training, lunged forcefully toward his sparring partner and felt a snap. The meniscus, a piece of cartilage that forms a cushion between the bones of his knee joint, had just torn. Soon he went in for surgery, undergoing a common, minimally invasive procedure in which the torn part of the meniscus is surgically removed. He recovered well, and both he and the surgeon were pleased with the outcome.

Some twenty-five years later, in 2007, a group of medical doctors in Helsinki, Raine Sihvonen and colleagues, asked a surprising question about the kind of surgery my father had. Did the surgery help recovery by repairing his torn cartilage, or was it all in his head? In other words, *did surgery help simply because patients knew they had surgery, and not because it helped repair the knee?*

To answer this seemingly absurd question, Sihvonen recruited 146 patients who needed surgery for a torn meniscus. At first, the surgeon made small incisions and inspected the knee joint for meniscus tears. If the surgeon found a tear, a nurse opened an envelope. Inside the envelope, a piece of paper instructed the surgeon either to go ahead and do the real surgery to remove the torn part of the meniscus, or else perform a sham surgery in which the surgeon went through all the motions of removing the torn meniscus, but *didn't actually remove it*. In other words, no surgical repair was performed. The patients had agreed to participate in the study and knew there was a possibility they would be in the sham group, but they were not told whether they actually had real or sham surgery. For the next year, the doctors followed the patients to see how their knee pain fared. The group that had the real surgery showed significant improvement, as expected. But then came the shocker: The sham surgery group also showed significant improvement in pain scores. Not only that, but *the sham surgery patients showed every bit as much improvement as the real surgery group at two, six, and twelve months after surgery*. This is despite the fact that meniscus surgery leads to improved outcomes as compared with not having surgery. As it turns out, my father's knowledge that he was having knee surgery for a torn meniscus might have contributed to most or all of the improvement he experienced from the surgery.

To understand why sham surgery may be effective, we need to explore the placebo effect. Despite being one of the best-known mind-body interactions, placebo effects are often spoken of dismissively, framed as "powerless" or "a myth." They are nonetheless very real. Perhaps some of the effects of prayer for healing could also be due to a placebo effect?

In the fourteenth century, before the placebo concept was associated with medicine, the word itself meant to flatter, often in conjunction with the verbs "to sing" or "to play." In Latin, it means "I will please." In "The Parson's Tale," which closes out *The Canterbury Tales*, Geoffrey Chaucer wrote that "flatterers are the devil's chaplains that sing ever placebo." The reference was to people who sang at funerals not out of grief for the deceased but to humor the bereaved because the flatterers wished to get fed.

In modern times, the word is most associated with relief of symptoms—usually but not always pain—by way of what is colloquially known as a sugar pill. But placebo effects continue to have a dubious reputation, as if they involve a transposition of the real and the imagined, or are simply a nuisance that muddies the waters of our scientific understanding of how therapies work. Researchers distinguish a placebo effect from a treatment that has a clearly understood pharmacological mechanism, even though a fake-pill placebo may ironically outperform a drug intended to provide the same symptom relief.

Placebos in the Brain

There have been many studies on how pain can be relieved via a placebo effect. For example, in one study people were given a mild but still unpleasant electric shock to the right wrist. In one condition, the researchers rubbed ordinary hand lotion onto their wrists and told them it was a special analgesic cream (i.e., a cream with pain-relieving properties) that would make the shocks hurt less. In the control condition, the researchers rubbed the exact same cream on the people's wrists but told them it was just ordinary cream and would not help alleviate the pain of the shocks. When they were told they were getting the special cream with pain-relieving properties, people rated the shocks 22 percent less painful. This kind of placebo response was reported by a stunning 72 percent of people in the experiment.

Such a significant result prompts the question: What was going on in the subjects' brains that made them feel less pain even when there was no difference in the pain-relief properties of the cream they received? To understand how placebos reduce pain, it helps to start by looking at where and how pain is represented in the brain, a question my own neuroscience research has asked. It turns out that the brains of those who experienced the placebo effect responded as if the pain was actually reduced. In other words, the pain-sensitive regions of their brains showed a significant decrease in pain-related activity.

The placebo effect isn't all in the mind, however. There's a true mind-body interaction, as there is a biochemical basis to the placebo effect. When we're given a placebo, we experience pain relief and/or the relief of psychological distress because our brains release endorphins and other chemicals that block pain-processing cells in the brain.

Many people have heard of, or have experienced, runner's high: the rush of euphoria and calm we feel after a workout. That high is not just "all in our heads"—it's *literally* in our heads. When we work out hard, the hypothalamus and pituitary in the brain release chemicals called endorphins. They block pain, lift mood, and help us keep going. Endorphins are a lot like powerful opiate painkillers such as morphine, heroin, or fentanyl, but they are all natural and made by our own bodies. We won't get addicted to them, even though we may get addicted to working out. Endorphins signal special receptor molecules in other parts of the brain, dialing down pain. So, when we are running, we feel amazing. Our muscles may hurt, but because of endorphins, the brain doesn't feel the pain. The next day the endorphins wear off, and our muscles might feel sore then. When we work out, it's like the pain receptor neurons in our muscles are on fire, and endorphins are like water spraying on the fire so we can keep going.

We know endorphins can block our pain as part of a placebo effect because, if we block endorphins themselves, some placebo effects stop working. In neuroscience, we often prove things in a way that may sound backward, as in the case above. For example, if we think that signal A (in this case, endorphins) causes signal B (pain relief), then, the conclusion goes, blocking signal A should also prevent signal B from being activated. So, we test the causation of A and B by blocking signal A. In this case, if we think the endorphins dampen the activity associated with pain, the question we'd ask is: If we block the endorphins, does that also prevent the pain relief from being activated? It turns out this is exactly what happens.

Many of us know someone who overdosed on fentanyl. When my daughter was five, the doctors gave her too much opiate painkiller, including fentanyl, during and after removing her tonsils. Once we brought her home, she became unresponsive and had to be rushed to the emergency room. Fortunately, there are treatments for opiate overdoses. Naloxone is a substance used in emergencies to save someone's life from a fentanyl overdose. Naloxone counteracts our cells' reception of any opiates. In other words, naloxone blocks the cells from responding to the opiates. This is a good thing, of course, for saving the life of someone who is overdosing. By blocking the cells' reception of fentanyl, naloxone reactivates the respiratory system and makes us want to breathe again. But naloxone blocks our cells from responding not only to misused opiates like fentanyl or heroin but also good opiates like naturally occurring endorphins. The recipient of the naloxone won't just want to breathe again. They also start to feel pain they hadn't felt under the influence of the opiates. By blocking the endorphins, naloxone dramatically weakens the pain relief they'd otherwise experience. This is where our test comes into play. By blocking A (the endorphins), we see the failure to activate B (the pain relief), showing the causation between A and B—indicating that the *endorphins cause the pain relief.*

Because endorphins cause pain relief, the theory might go, placebos will also relieve pain if placebos trigger that endorphin release. And that's what studies show: Effective placebos increase endorphin levels, and these can also be blocked by naloxone, thus blocking the pain-relieving effects of placebos.

When I was a child and scraped my knee, my mother would put a bandage on it and tell me the bandage would make my injury feel better. On the face of it, that bandage, even if it was a super-special one with a superhero on it, was doing nothing of the sort. It was there to protect the wound, period. But, like many kids before me, once my mother told me the bandage would help, I immediately felt better. The hug and attention I got from my mother had more to do with my symptom relief than the bandage, of course. Also, her authority mattered. I believed what she said, and that belief led me to expect that the bandage, though it also could have been a sugar pill, would work. This positive belief triggered an endorphin release, which made me feel better. The hug, the attention, and her authority all resulted in my expectation of relief, and it's the expectation of relief that triggers the endorphin release and makes placebos effective.

It turns out that there are multiple ways that placebo effects can reduce pain, not only by endorphins. Much like ibuprofen and morphine block pain with different chemical effects on the body, our bodies can also release different kinds of placebo chemicals to reduce pain, not just endorphins. One of those ways involves chemicals called cannabinoids, which are found in marijuana and can relieve pain, but which our bodies also produce naturally. These are called *endocannabinoids*. The first clue in understanding how they work is found with a drug called rimonabant. Just like naloxone blocks pain relief from opioids, rimonabant blocks pain relief from marijuana. It does this by blocking a receptor molecule in our bodies called CB1. Remarkably, the same pain-relieving effects of stimulating CB1 receptors can be induced by placebos.

Fabrizio Benedetti, Professor of Physiology and Neuroscience at the University of Turin Medical School in Italy, used rimonabant in 2011 to show that the placebo effect could relieve pain even without endorphins. To do this, they gave human participants a series of pain tolerance tests over five days. One group was given morphine (an opioid) on days two and three, and then a placebo that they thought was a pain reliever on day four. Subjects experienced pain relief from the placebo on day four during the pain tolerance test, and rimonabant did not interfere with the placebo pain relief. This finding showed that for those subjects, their body had learned to expect opioids. Moreover, their bodies were producing more of their own endorphins to help meet their expectation of pain relief from the morphine, an opioid. Another group, on days two and three, was given ketocorolac instead of morphine. Ketocorolac is a pain reliever with the same mechanism of action as ibuprofen, but more powerful. It acts on CB1 receptors to alleviate pain. This time, on day four when they weren't given any ketocorolac but expected to get it, subjects would have experienced a placebo effect of pain relief, but the rimonabant prevented the pain relief from working. This demonstrated that the placebo effect produced by prior experience with ketocorolac was working by activating their CB1 receptors, and it completely bypassed the endorphin mechanism of placebo pain relief. This research shows that the human body has multiple complex biochemical pathways by which placebo effects can reduce pain. *There is not just one placebo effect—there are many.*

Placebos for More than Pain

Some of these placebo effects can alleviate not only pain but depression as well. In a series of studies of placebo pills given as antidepressants, the depressed patients' symptoms were reduced by 52 percent with antidepressant medication but 34 percent with placebo alone. How could a placebo be almost as effective as the antidepressant meds? Think of the millions of prescriptions for Prozac and Zoloft and the like. These drugs are selective serotonin reuptake inhibitors (SSRIs), which relieve depression by enhancing the effects of the neurotransmitter serotonin in the brain. It might seem incredible that nothing but sugar pills could substantially relieve depression symptoms, but in some ways it's not so surprising. One way to think about it is to consider all the care the placebo-treated patients received in the cited studies. They each had a thorough psychiatric and medical evaluation. They probably experienced immense relief at having a committed and enthusiastic medical expert diagnose their distress. A treatment plan developed by experts gave them renewed confidence that help was on the way for their situation. And finally, but no less important, they had the opportunity to verbalize their distress. As with my mom and the bandage, care and attention to one's wounds and a reinforcing statement from a medical authority can lead to positive belief, and thus: a placebo effect. In fact, studies show that placebos can be almost as powerful as a pharmaceutical treatment like an SSRI.

The placebo effect isn't even limited to pain and depression. There is evidence that placebo effects can improve symptoms across a range of illnesses. For example, Kevin Olden, a practicing gastroenterologist, and colleagues in 2002 tested the drug alosetron as a treatment for irritable bowel syndrome (IBS). IBS is a chronic condition that can cause significant distress, with discomfort and frequent trips to the bathroom substantially impairing a patient's quality of life. The researchers gave women with IBS either alosetron or a placebo pill daily for twelve weeks, then asked them how satisfied they were with the relief provided by the medication. Prior to the study, less than 10 percent of patients reported they were satisfied with their current IBS medication. After twelve weeks in the study, 69 percent of those taking alosetron reported being "satisfied" or "very satisfied" with the results of their treatment. Remarkably, 46 percent of those taking the placebo pill were also "satisfied" or "very satisfied" with how effectively the medication controlled their symptoms. In other words, the placebo effect accounted for about two-thirds of the medication effect.

Perhaps more surprisingly, a placebo effect can also improve symptoms of diseases such as Parkinson's. This is an incurable, progressive disease in which the dopamine cells in the brain die, and patients have increasing difficulty moving their body and thinking clearly until they eventually succumb. There are medications like levodopa and carbidopa that can help alleviate symptoms, but most patients with Parkinson's experience "off" times, when their regular medication wears off and they again have difficulty moving due to the disease, while they wait until they can take the next dose. Selegiline is a pill that can help reduce the length of off times, thereby improving patients' quality of life. William Ondo, director of the Movement Disorders Clinic at the Houston Methodist Neurological Institute, tested how well selegiline can reduce off times over the course of twelve weeks, compared to a placebo pill. After twelve weeks, the off time was reduced by 12 percent in those taking selegiline, and 10 percent in those taking the placebo.

In this study, simply believing that they were taking an effective drug was enough to improve patient symptoms almost as much as the active drug itself. In context, most of the symptoms persisted, and the improvement amounted to better movement ability for just a few minutes per day, so the placebo effect was by no means a panacea. Nevertheless, the improvement from the placebo effect was still significant. Another study showed that not only did the Parkinson's disease symptoms improve, but the placebo actually changed the brain activation, just as if the patient had received regular treatment with deep brain stimulation, an effective but invasive surgical therapy.

There are still more studies that show placebo effects for other conditions such as ulcerative colitis and high blood pressure. There's not a lot of evidence for placebo effects and cancer improvement specifically, but placebo effects address both pain and depression, which can otherwise shorten cancer survival times. Either way, there's clear and consistent evidence that, overall, placebo effects work and that they alleviate unpleasant symptoms.

Despite the demonstrated efficacy of placebos, there's still remarkable resistance in both religious and nonreligious communities to accepting prayer as a therapeutically beneficial treatment that may be partially explained by placebo effects. For scientists, the fact that there's no proven pharmaceutical active ingredient in the placebo puts it in the category of quackery. The stain of deception still hangs heavy over the idea that placebos can be an effective treatment for everything from pain to depression and more. And for many religious people, explaining effects of healing prayer as a placebo is seen as a lack of faith; they may bristle at the idea that the healing they or others experience may be all in their heads.

To say something has a placebo effect is often to dismiss it. But why dismiss something that is clearly helping us? If anything, we should be studying how placebo and prayer can work in tandem. It's not an either/or. And perhaps surprisingly, *placebo effects can relieve symptoms even if the person receiving them knows that they are taking a placebo*! Linda Buonanno had been suffering from IBS for decades when she signed up for a clinical trial of a new treatment. Once she enrolled, she was told that she would definitely be given a placebo pill—it had no active medical ingredients whatsoever. She was disappointed but decided to try it anyway. To her surprise, her symptoms vanished, for the first time in decades. The study ended, and according to *Time* magazine, "her doctors . . . tell her she can probably stop taking the placebo pills and that it's unlikely she would have any problems, but after several recurrences, she is too scared to do so." Buonanno's experience was not unique. Other IBS patients in the same study, which was conducted by Ted Kaptchuk and colleagues at Harvard Medical School, were all told explicitly that they were being given placebo pills with no active drug. Those taking the placebo pills had significantly more symptom improvement than those who took no pills but had the same doctor visits. Even if someone knows they're taking a placebo that has no active ingredients, *the key for making it work is believing it might help*. We may be more likely to believe in the placebo's, or any treatment's, efficacy if we are taking advice from a trusted authority (a bandage-wielding mom or a medical doctor), and you believe that following their advice may help you. That belief alone can change your body, causing as yet unknown chemical changes that can significantly alleviate symptoms.

In chapter 3, we considered that ideally, we would hold beliefs that are both empirically true and good for our health. As we see now, placebo effects constitute a tautology in this regard, a kind of self-fulfilling prophecy: We believe they are effective because they work, and they work because we believe they are effective. While in the most ideal world our beliefs would be objectively true, evidence from placebo effects suggests an oddly reassuring if unsettling conclusion: *The objective truth of our beliefs is not always a prerequisite for experiencing the health benefits they provide.*

The clear symptom relief that the placebo effect provides over a wide range of illnesses also raises the question of how many diseases have a mind-body connection to them. To make a tumor disappear might be a higher-magnitude cure, but if specific types of cancer *are* affected by our mental state or stress, then perhaps therapies that take this interaction into account can, at the very least, help with symptom relief. We might not know which illnesses placebos can or can't help or even all the biochemical effects of placebos. We also might not know which illnesses prayers can or can't help. But shouldn't all this be a larger topic of investigation? As we'll see in the next chapter, beliefs can have even more powerful and direct mind-body effects on our health than placebo effects.

8

Mind-Body Effects

ON FEBRUARY 29, 2004, ABOUT SIX MONTHS AFTER I WAS DIagnosed with a glioma, I found myself standing in a church in Santiago, Cuba, not far from the US military base at Guantanamo Bay. I had come to investigate reports of remarkable, even miraculous healings in Cuba. My curiosity was more than intellectual: I hoped to find something that could help heal my brain tumor and was willing to travel, since medically I had no hope of a cure. The team of about twenty others I came with were part of a Christian organization that regularly prayed for healing and claimed to see miraculous results. The church we were visiting that day was in an urban area near our hotel. The building was fairly large, mostly made of brick and concrete, with palm trees outside. Inside were wooden pews and white tile floors that are common in churches in Latin America. About three hundred people were present for the Sunday evening service. As they prayed, and as members of our group began to tell stories of dramatic healings they had recently seen, the people in the congregation seemed eager to receive healing prayer for themselves.

We divided into teams, often two of us plus a local who could translate. A new friend on the team, Doug, and I prayed with a woman, whom I'll call Cristina. She was in her late twenties or early thirties and had lost her eyesight two or three years before. Cristina told us that since her sudden vision loss, she couldn't even see a hand waving directly in front of her face. She told us that she wanted to be able to see again, and she asked us to pray for her healing.

We three stood together on the stark white floor near the front of the church. As Doug and I faced Cristina, we began to pray. I was still new to the idea of praying for healing and wasn't quite sure what I was supposed to be doing, so I repeated the same simple prayers I heard other people praying. We alternately asked God to heal Cristina's eyes and then, still speaking gently, commanded her eyes to see on the authority of Jesus. After a few minutes, we asked Cristina if she noticed any difference in her eyesight. She did not. We continued to pray for about thirty minutes, but Cristina still reported no improvement whatsoever. It was discouraging. People all around us were excitedly talking about various healings and improvements that their prayers had prompted. I began to wonder if there was a technique that Doug and I were missing.

Finally, at Doug's urging (since I knew enough Spanish to translate for him), I asked Cristina whether something had happened just prior to the time when she lost her eyesight.

"No, nothing," Cristina said.

We pressed her. "Are you sure nothing traumatic happened around the time of your vision loss?"

"No."

After a brief pause, Cristina seemed suddenly to make a connection.

"Now that I think about it, maybe there was," she said slowly. "One day the police came suddenly and arrested my husband. They imprisoned him, and I haven't seen him since. That was right before I lost my eyesight. I never put the two together, but is that the kind of event you mean?"

"That sounds very traumatic," I responded. I couldn't imagine it if the police had come and imprisoned Candy. What must Cristina be going through?

"Yes, it was very sudden. After they took my husband away, I was a wreck, fearful and anxious all the time. I had no reason to believe that I was under investigation, but I'm still not sure if or when I will ever see him again." Cristina was beginning to tear up.

"I'm so very sorry," I said.

"I've been so worried. My family has been helping me. They've been great, but still, without my husband, I feel very alone."

"You've been through a lot," I said.

By this point Cristina was really crying. Doug and I did the best we could to comfort her. We hugged her and let her know our concern for how much pain she must be in. Doug and I then decided to shift our approach to praying for Cristina, focusing on healing her emotional pain, especially her fear and anxiety, rather than her eyesight.

After a few minutes of our praying, Cristina managed a smile through her tears. Doug and I then decided to pray again for her eyesight to be healed. The three of us were standing in the middle of the church. Cristina stood between Doug and me as we faced each other and each gently laid a hand on top of her shoulder. I began to pray, "Jesus . . . ," but I only got halfway through the first word. Instantly, I felt an incredibly strong sensation, like a burst of electricity all through my body. Doug later said that he felt the same sensation at the exact same time. Cristina opened her eyes, blinked several times, and then screamed in Spanish, "*Soy sana*. I'm healed!"

Doug and I looked at each other in surprise. Although I had prayed with others for healing on occasion as many Christians do when faced with illness, I had never seen such a dramatic healing that immediately followed my prayers for someone. I couldn't think of what to say, so I just hugged Cristina.

"Soy sana, soy sana," Cristina exclaimed with joy, as she returned my hug.

"What can you do now that you couldn't do before?" I asked.

It was one of the standard questions the group asked. Partly, we wanted to know whether people reporting healing had experienced relief from all their symptoms or more limited improvement, in which case we would offer to continue praying. Partly, it was simple curiosity. When people tell us they are healed, we want to know what's changed.

I held up a hand in front of Cristina's face.

"How many fingers do you see?" I asked, like an ad hoc optometrist.

"Three," Cristina replied. She'd counted correctly.

"And now?"

"Two."

"That's correct!"

I then moved farther back. Cristina continued to answer correctly. Finally, I stood all the way across the church building, while Cristina looked at me.

"How many now?" I asked.

"Four!" she said triumphantly.

It was incredible. Cristina was easily counting the fingers I held up, even from across the building. It did indeed seem as if she could see clearly for the first time in several years. The tears and look of sheer joy on Cristina's face were beautiful.

I was in shock walking back to my hotel room later that night. I had witnessed a person's eyesight instantly restored, right in front of me. It seemed unlikely that anyone had orchestrated an elaborate fabrication, especially on the fly, given that we had first stumbled across the church hours earlier. Of course, I'd be naive not to acknowledge that individuals may have reasons for fabricating or lying, whether for attention, money, or influence. All I can say is, that's not what seemed to have happened here. But what happened, exactly? What part of Cristina's eyes had been damaged to start with and what part had been healed? I hadn't noticed any visible damage to the lens of her eye or even any discoloration in the pupil, but I'm not an optometrist nor ophthalmologist—it's possible I missed something. So, I couldn't say one way or another what the medical issue she'd had was. It was possible that Cristina had had what neurologists call a *conversion disorder*.

Conversion Disorder

It is well known in the scientific and medical communities that certain cases of blindness can be caused by psychological conflicts or stressors. These cases often resolve spontaneously as the stressor diminishes. These so-called conversion disorders are particularly common in women ages ten to thirty-five. Conversion disorders usually involve a visual impairment or some kind of musculoskeletal weakness or paralysis. The symptoms can include muscle spasms or twitches, pain, and seizures or fainting. The disorders are often anxiety-related or triggered by a stressful event. In the nineteenth century, Jean-Martin Charcot, a mentor to Sigmund Freud, associated conversion disorder with hysteria. Charcot designated the symptoms described above as functional, meaning there was no visible cause nor was there any evidence of organic (i.e., physical) disease. Freud built on Charcot's theories to say that patients with conversion disorders had likely had some traumatic experience that had been repressed or pushed into the unconscious. Their psychological distress symptoms were then "converted" into physical symptoms.

In cases of visual impairment, the diagnosis of conversion disorder is often given when there is a loss of vision despite no evidence of damage to the visual system. In most patients, symptoms resolve within a week, and patients regain normal eyesight. In some patients, however, symptoms may recur or become chronic. Treatment typically involves medication and behavioral therapy to alleviate anxiety, depression, and related concerns, and physical therapy if the condition involves weakness or paralysis. It is important to note that patients with conversion disorder are not intentionally faking their symptoms—they are genuinely experiencing the impairments.

Cases of conversion disorder causing visual limitations are not unique. In one case from 2014, a twenty-three-year-old woman was brought to the emergency room after she passed out for a few seconds. She reported experiencing weakness in one leg and blurry vision. While she was still in the ER, her condition developed into a loss of vision in both eyes. Within a few hours she reported that she could only see faint shadows. After comprehensive medical tests, including neurological tests, CT, and MRI scans, no cause could be found for the vision loss. When the doctor inquired about her history, it turned out she was experiencing psychological stress—she worked two jobs, attended school, struggled financially, and was a single parent of a four-year-old child, all while dealing with difficulties involving the child's father.

Immediately after being diagnosed, this woman was given a brief cognitive behavioral therapy (CBT) intervention. She was then referred to a psychiatrist, who provided her with outpatient psychological support. By the next day, she felt better psychologically. Her vision improved somewhat, but it was still blurry. Over the next few days, however, her symptoms gradually disappeared. By a week later, her blindness was cured, and she was in complete remission.

So, what happened to Cristina when she regained her eyesight during our prayer? Did the absence of obvious ocular damage or cataracts suggest conversion disorder? My team members, all of whom believed that God heals supernaturally, immediately called it a miracle. For the absolute naturalist, the only possibility is that Cristina had a conversion disorder, especially given her traumatic history, and that she experienced healing as her psychological trauma was acknowledged, embraced, and alleviated. As a methodological naturalist, I would ordinarily assume that Cristina had a conversion disorder. Yet Cristina's recovery was nonetheless remarkable given how quickly it happened and how much time had elapsed since she lost her vision in the first place.

If healing miracles do sometimes happen—if the supernatural exists and occasionally intervenes in an otherwise natural world—would it look any different from this? Without better examination equipment or medical records, we cannot be certain what happened to Cristina neurologically. If it was a conversion disorder, it would have been difficult to verify since we don't understand how they work biochemically. Whatever the case, she was overjoyed to be able to see again, and her emotional pain had also been alleviated.

Defining Miracles

Cristina's healing might seem, at least to some, an ideal example of a miracle. But in my mind, it raises the question of what exactly *is* a miracle, or more precisely, what are the criteria for calling something a miracle? I distinguish here between calling something a miracle versus claiming that it must have a supernatural origin. A miracle is an event we can define by a set of criteria. If the event meets the criteria, then we can call it a miracle. But supernatural causation is an inference—and that I leave to the theologians to debate. The word "miracle," however, is tossed around so loosely in everyday speech that its definition often appears elusive: "I found a parking space at the shopping mall on the weekend before Christmas—it's a miracle!" If everything is a miracle, then nothing is a miracle.

The Roman Catholic Church, as it turns out, has been working with a definition of miracles since at least the seventeenth century. Cardinal Prospero Lorenzo Lambertini (1675–1758), who later became Pope Benedict XIV, had a long-standing interest in science. The Catholic Church by that time had a well-developed tradition of recognizing saints—persons in heaven who lived heroically virtuous lives, offered their lives for others, or suffered martyrdom for their faith, and who are worthy of imitation.

The Church has an official procedure for proclaiming someone a saint—and that procedure requires miracles. Canonization of a new saint typically involves *proving* that the intercessory prayers of a deceased person in heaven resulted in two miracles, some 95 percent of which have, in the history of canonization proceedings, involved physical healing. Over time, the Church developed formal procedures for investigating alleged miracles. Think about how US courts of law operate—it's very similar in the Catholic Church. For much of the Church's history, a Promoter of the Cause, or *advocatus Dei* ("God's advocate"), presented evidence, including from medical experts, that a miracle had occurred, generally in response to petitions for the candidate's intercessory prayer. A Promoter of the Faith, or *advocatus diaboli* ("devil's advocate"), looked for weaknesses in the evidence. The devil's advocate proposed natural explanations for supposed miracles, thus preventing the candidate from being officially declared a saint. Although the 1983 *Code of Canon Law* reduced the role played by a specific devil's advocate, canonization proceedings still include skeptical evaluation of evidence and systematic efforts to *disprove* that healings following petitions for intercessory prayer require a miraculous explanation.

Cardinal Lambertini served as devil's advocate from 1708 to 1728, giving him an unusual opportunity to consider alternative explanations for alleged miracles. He was a devoted student of human anatomy with extensive knowledge of medical literature. Lambertini was not the first Church leader to propose tests for identifying miracles. But he systematized rigorous criteria for differentiating healings that might be explained naturalistically from miracles that could not be disproven.

Lambertini brought to his role as devil's advocate a healthy degree of skepticism. As he saw it, recovery from diseases that might otherwise resolve themselves, such as certain psychological conditions, should not be declared a miracle for the purpose of canonization. He thus ruled out what we would today consider psychosomatic or conversion disorders. Lambertini was also suspicious of epilepsy cures, noting that seizures might go into remission but come back after some time; their disappearance thus did not constitute a miracle. In one canonization proceeding, Lambertini checked back after eighteen months to ensure that the patient with epilepsy was still in remission before designating the cure as a miracle.

The standards used by the cardinal are now known as the Lambertini criteria. To be declared an official miracle, for the purpose of canonization, a healing had to pass five tests. First, the disease must be serious and at least difficult if not impossible to cure by medical means. Second, the recovery must be sudden and instantaneous. Third, the recovery must be complete. Fourth, the recovery must be permanent. Fifth, it must be proven that the recovery cannot be explained by medical treatment or the healing power of nature. The point of these criteria is that a saint should not be canonized lightly—the Church seeks a high degree of confidence that a miracle happened, and that claims are neither fraudulent nor simply unconvincing upon critical inquiry.

Today, the Vatican still follows basically the same process to examine miracles, hiring medical experts to testify on the merits of a miracle case. Having heard the case, a committee of the Dicastery for the Causes of Saints votes on the matter, and the committee's recommendation is sent to the pope for final approval. Provided that the Lambertini criteria are met, the Roman Catholic Church will formally label the healing as a miracle.

Jacalyn Duffin is a historian and medical doctor with a specialization in hematology who studied original records of miracle claims in the Vatican's Secret Archives. In her book *Medical Miracles*, Duffin describes her research on miracles and how her interest began. She was asked in 1987 to read bone marrow samples from a leukemia patient who was, in her expert opinion, inevitably going to die soon, but who nevertheless survived. The samples, unbeknownst to her at the time, were provided by a Catholic organization investigating the case as a potential miracle. Duffin was later invited to the canonization ceremony, and at that point she was hooked.

Duffin began researching miracles in the Vatican archives asking whether healings declared miraculous in the past could withstand modern medical scrutiny. She came away from her research with two conclusions: First, as far as her own personal religious beliefs, she remained an atheist; and, second, she now believed in miracles. "These events were miracles for the people involved," she affirmed. The rigor of Catholic investigations and willingness to set aside cases falling short of the Lambertini criteria impressed her. Thus, in specifying that events were miracles "for the people involved," I don't think Duffin meant to dismiss belief in miracles as inherently credulous.

Rather, Duffin was saying something profound about the nature of belief—whether in the supernatural or absolute naturalism. In Duffin's words, "neither God nor the elusive and as-yet-unknown natural explanations, which my medical colleagues are convinced must exist, can be falsified. . . . Both are beliefs, and they fall outside the realm of the scientific method as we know it." Duffin adds that "because the one belief"—absolute naturalism—"utterly pervades the scientific community, it seems not to be a belief but a 'fact.'" As Duffin saw it, for her absolutely naturalistic colleagues, no amount or type of evidence for the miraculous could ever be persuasive. Preexisting *belief* that every healing must have a natural cause led to their refusal to even entertain the possibility of supernatural causation. In a sense, finding a natural cause for a previously unexplainable event would indeed falsify the claim that an event could only have been supernaturally caused. Here, Duffin seems not to be referring so much to her colleagues accepting the possibility that a natural cause could be found, which is reasonable, but to an absolute faith that such a cause *must* exist, even if it is never found. This fits with ideas we've been discussing about cognitive preconceptions like confirmation bias that can make it difficult for us to evaluate evidence—whether for miracles or other controversial claims—objectively.

Setting aside for now the question of whether miracles could ever prove God's existence or intervention in the natural world, let's ask a more pragmatic question: What can healings experienced as miracles tell us about what is possible for our own healing, especially when doctors offer no hope?

My friend Dirk Kruijthoff is a medical doctor in the Netherlands who stumbled on one such seemingly miraculous case and decided to explore what is possible. His findings offer insight into what can happen medically in cases of healing after prayer. A patient of his in Amsterdam suddenly experienced healing during a prayer meeting in 2007. The patient had been confined to a wheelchair with post-traumatic dystrophy for eighteen years following a series of minor injuries and surgeries. After receiving prayer for healing, the pain in her feet left instantly and permanently, to the point that she no longer required her usual doses of heavy opioid painkillers. Kruijthoff told me that he was rather skeptical of such claims before 2007, but since he knew his patient's medical details so well, he felt the case was compelling and decided to do a PhD project investigating similar cases. In the end, he and his research team considered 83 cases of healing through prayer, largely in Europe. Kruijthoff and his medical colleagues applied the Lambertini criteria to 27 of the cases, concluding that 11 cases were "medically remarkable" but that none could be considered "medically unexplained" in a way that would meet the Lambertini criteria.

One of the medically remarkable cases was Corlien's, a woman with Parkinson's disease. In 2009, at age fifty, Corlien experienced stiffness on her right side, with difficulty in writing, and she could not make normal facial expressions. She was diagnosed and given medication for Parkinson's disease. Her symptoms worsened, and though she took heavier doses of medication, it started wearing off before she could take the next one. She needed a wheelchair much of the time and had difficulty even carrying on a conversation. On April 6, 2012, three years after she was diagnosed, she had an unexpected healing prayer experience at a church. Immediately afterward, most of her symptoms—except for slight stiffness—disappeared, and she stopped taking most of her medication. She remained well for at least eight years.

This raises the question: Was Corlien's brain miraculously healed? The short answer is no, at least according to available medical tests. Three years after her remarkable recovery, doctors scanned Corlien's brain with a special machine (DaT scan) that measures dopamine levels. If her brain had been objectively restored, the dopamine levels would have returned to normal, but they did not. The brain scan showed she still had Parkinson's disease, even though her symptoms were almost completely resolved and had been for years. While a placebo effect cannot be ruled out, the recovery was faster and greater than what is usually seen with placebo effects. If the scan had indicated healing of Corlien's brain, Kruijthoff might have concluded that her case met the Lambertini criteria and called it a miracle. But because the brain still showed damage, the team labeled the case remarkable, falling just short of inexplicable.

Corlien's case suggests that *clinical recovery from incurable diseases is possible, even beyond what placebo effects normally accomplish*. The case also suggests that people may experience healing as their symptoms disappear—even if the disease itself still appears on medical tests. Corlien's doctors gave her no hope of a cure, and they would certainly not have expected her brain to show disease while her symptoms resolved. As Corlien put it, "my body does not fit in your medical textbooks."

Remarkable healings, or at least remarkable resolutions of symptoms, are difficult to study because they are by nature rare. Certain places around the world are particularly known for claims of miracles, such as the shrine in Lourdes in France, but it would be difficult to study apparent miracles in process there because of their rarity. Just seventy-two cases of healing at Lourdes have been declared miracles by the Catholic Church since 1883, or barely one miracle every two years. Miracle claims are much more common in some lesser-known areas, such as Mozambique and Brazil. We turn to research in these regions next to answer the question: Can we measure a remarkable healing, or even a miracle, while it is in progress? And if so, what can we learn about how such healing works?

9

Chasing Miracles

A HOT AND HUMID SEA BREEZE DRIFTED THROUGH THE BEAUtiful mango trees. Candy and I had just unpacked in a picturesque stone guest house that was only a short walk from the ocean. Moments before, we'd driven in a jeep down a dirt road past a row of thatched roof buildings. Everywhere we were surrounded by green palms, an azure-blue sky, and rich red dirt. It was June 2009, and we had just arrived for a two-week stay at the Iris Ministries base at Pemba, in Mozambique's embattled north, a beautiful land filled with civil unrest. Iris, like many nongovernmental organizations, offers aid to families, individuals with special needs, and widows and orphaned children displaced by the conflict. Iris provides daily lunch for local children, many of whom have found protection from the region's instability within the compound's high stone walls, and their medical center offers free care to those who lack resources. Beyond their base, Iris helps build churches, schools, wells, and health clinics in neighboring communities.

Iris Ministries and its leader, Heidi Baker, regularly held meetings that included prayer for healing. Heidi had finished her PhD in systematic theology from King's College London in 1995, after which she and her husband, Rolland Baker, moved to Mozambique to focus on caring for children in an orphanage. Over time her missionary work expanded to include evangelistic outreaches in addition to caring for orphaned children. Reportedly, over the last two years, most of the people who asked for healing of deafness, and many who wanted healing of blindness, had their prayers answered. We greeted the news of these reported healings with a mix of academic caution, curiosity, and grave excitement. Now we had come to Mozambique with a specific goal of studying remarkable healings through prayer as they were happening. We had several questions: What did healing prayer look like, and how did it work? Were people indeed claiming to experience healing of blindness and deafness? What was happening to them, medically speaking? We brought along diagnostic equipment, including vision charts and an audiometer, that would allow us to measure vision and hearing. Our plan was to administer vision and hearing tests immediately before and then immediately after the healing prayers, to measure any changes, and then correlate those changes with specific aspects of how the healing prayers were conducted. This would help us catch a miracle in the act, in a manner of speaking. My approach, as is always the goal of science, was to investigate, predict, and explain, and through testing determine the reliability of those explanations.

My wife led the project. As a historian, ethnographer, and professor of religious studies at Indiana University, with her bachelor's, master's, and doctor of philosophy degrees from Harvard, Candy—whom I have cited throughout this book—was well qualified to study religious practices. After I was diagnosed with a brain tumor, she began studying spiritual healing practices as part academic interest and part personal quest. She eventually wrote her own book, *Testing Prayer*, published by Harvard University Press in 2012. For this work, Candy had received funding from the John Templeton Foundation for our travel and research expenses. We consulted with colleagues who had the necessary medical expertise—including specialists in otolaryngology, an ophthalmologist, and a professor of optometry—to help us design the study and carry it out properly.

Pemba, the capital of northern Mozambique's Cabo Delgado province, where Iris Ministries was headquartered, seemed a promising location to conduct our research on healing prayer. Pemba has little in the way of Western medicine available, and people there have strong belief in healing through spiritual power—much more so than in the West. If belief can influence health outcomes, we would expect to see larger effects of prayer for healing where the foundations of belief are strong.

The health conditions we focused on, vision and hearing impairment, were easier to measure than many others. With, for example, back pain, it's harder to come up with a specific measurement of better and worse. You're reliant on a patient's self-report of pain. But with vision impairment, there are well-developed tests that tell us whether someone has 20/20 vision or 20/40 vision. Similarly, there are standard tests for the degree of hearing loss an individual has. We brought with us modified Snellen eye charts, the kind found in any optometrist's office, except that all the letters were the capital letter "E." Moreover, the "E" appeared rotated so that the open end could face right (E), left (Ǝ), up, or down. Our participants could read the eye chart by indicating which way the "E" pointed. In this way, we could measure vision even in the part of the population that was illiterate. Likewise, we brought along a portable audiometer, which generated precisely controlled sounds of a specific volume and pitch administered through headphones. This allowed us to measure what was the faintest sound a person could hear.

We were still reliant on self-reporting, and that could skew the results. As Candy described extensively in *Testing Prayer*, we could not absolutely rule out some effects that might confound the results, such as holdback effects, where you purposefully sandbag or underperform in an initial test to make your improvement later seem greater. For example, if someone wanted to deceive us into thinking they had improved greatly after prayer, they could pretend to be blind or deaf during the before-prayer measurements. We addressed this to some extent by repeating measures to ensure they were consistent, reasoning that it is harder to fake a specific and consistent level of hearing loss, for example. Also, we wouldn't be able to access individuals' medical histories, something we would take measures to address in the future, but had no means of doing in this instance.

Early in the morning on June 9, 2009, Heidi and her team loaded up some trucks, and together we drove several hours, first over paved and then bumpy dirt roads, to several remote villages, the first of which was Namuno. The villages looked much as they probably did five hundred years ago, except for the presence of cars and old clamshell cell phones. The houses were made of mud walls and thatched roofs. There was no running water and almost no electricity. When evening came, Heidi set up a movie screen on a flatbed truck in the center of the village, where she proceeded to show a film about the life of Jesus, translated into the local Makua dialect. Generally, most of the village would show up, anywhere from a hundred to a thousand people, to watch the film. After that, Mama Aida, as people knew Baker, got up on the flatbed truck and said in half Portuguese, half Makua, like an actor who's taken the stage, "Where are the blind and deaf? Bring me all the blind and deaf in the village. Jesus will heal them." Baker, like many others who pray for healing, believes that the kingdom of heaven on earth actively opposes the forces of sickness and oppression. Her role, as she saw it, was to mediate the advancement of the kingdom of heaven, help free others from sickness and oppression, and thus heal them.

It was fairly dark in the village, but the generator-powered lights around the flatbed truck provided an island of light. About six people had lined up beside the stage, all of whom claimed to be deaf or at least severely hard of hearing.

Heidi stepped down from the stage to the side, where the hard of hearing people stood. I got out the audiometry equipment, and Candy and I placed headphones on each of the six subjects and asked them to indicate when they heard a tone, gradually increasing and decreasing the volume to find a consistent threshold at which they could hear. We repeated the process in each ear separately. This method is the same as what we'd undergo for a hearing test in an audiologist's office in the US. Some subjects could hear only extremely loud sounds, while some were unable to hear even the loudest sounds.

After that, it was Heidi's turn. She went down the line one person at a time, and hugged each person, smiling as she put her hands on each ear and commanded their ears to hear properly on the authority of Jesus. "God bless you!" Heidi would say to each one as she moved on to the next.

After Heidi had finished, I went down the line again, carefully repeating the measurements I had made only a few minutes earlier. Subjects experienced, on average, 20 decibels of improvement. One showed closer to 60 decibels of hearing threshold improvement, or the equivalent of going from barely hearing a motorcycle engine a few feet away to hearing normally, given the background noise. I also measured the background noise from the crowd to make sure that the apparently improved hearing wasn't simply due to the noise dying down. We carefully wrote down the measurement numbers, and when the meeting ended late at night, we pitched our camping tents in someone's backyard and slept.

The following morning, Heidi helped heat up some breakfast over an open campfire, and we all sat around chatting before breaking camp and moving to another village. The next few days played out similarly, with healing in the evening, sleeping in a tent, and sitting around a campfire in the morning.

Another night, by a similar flatbed truck–turned-stage in another village, Heidi called for the blind people to come up. A woman came up, whom I'll call Maryam, accompanied by her friends from the village who affirmed that she was blind. Her right eye was notably white. I held up one, or three, or four fingers about a foot in front of her face, asking if she could tell me how many fingers I was raising.

"I don't see. I can't see," Maryam said, staring ahead with both eyes open.

With that, Heidi gave her a hug and prayed briefly for her eyesight to be healed. After a few moments, Heidi asked for the near-vision eyechart. I stood next to Heidi and Maryam as Heidi began pointing at successively smaller "E" symbols and asking the woman to identify the symbol, while I made sure the eye chart was kept the appropriate distance from Maryam's eyes. Maryam pointed up, down, left, or right accurately now according to the symbol, as a big smile appeared on her face. She got down to the 20/125 line. The 20/20 line is normal vision, and the 20/125 line has somewhat larger letters. It is what a person with normal vision could see from 125 feet away, but a person with 20/125 vision would have to be not more than 20 feet away to see. Maryam seemed to have gone from being unable to see even a hand in front of her face to being able to identify relatively small symbols, all in the span of a few minutes.

Maryam's right eye remained visibly white and unchanged. I don't know how much of her vision the white area blocked. The opacity of her cornea could have been caused by trachoma, which is a germ related to chlamydia. Trachoma eventually scars the insides of the eyelids and causes the eyelids to turn inward. Over time, the inward-turned eyelids scratch the cornea, and if left untreated, the cornea can scar, leading to blindness. Or her eyes could have had a different disease. Without medical records, I can't say definitively one way or the other what caused her blindness. Either way, whether it was trachoma and/or something else, given that her eye remained white, knowing the specific diagnosis still wouldn't have explained how she regained sight. We were constrained for time and unable to test monocularly. That is, we didn't test the left and right eye separately, so there is a chance that the improvement she experienced was in her left eye, not her right eye. But either way, an improvement from blindness to the 20/125 line on an eye chart is significant.

Afterward, we analyzed the data in both the vision-impaired and hard of hearing individuals and found that the improvements were statistically significant across the population. In both cases, I had double-checked the measurements. Later, we replicated the study in Brazil, finding similar dramatic improvements in healing prayer meetings there. The effects were real, but the question remained: What could account for the effects?

Hypnosis and Suggestion

Hypnosis and the related concept of suggestion have been studied as treatments for a variety of health concerns since the eighteenth century. In 1759, Franz Mesmer was a medical doctor in Vienna who came to believe that he could heal patients by manipulating invisible energy in their bodies with magnets, a phenomenon he eventually called animal magnetism. Mesmer first applied his method in 1774 to treat a young woman who was reportedly cured of her otherwise intractable ailments, including fever, pain in the ears and teeth, and vomiting. In 1778, Mesmer moved to Paris, where he continued his practice, and many of his patients also believed themselves cured. (Mesmer's treatments would eventually introduce the verb "mesmerized" into the English lexicon.) But academics were skeptical, and by 1784, King Louis XVI appointed an inquiry panel that included Benjamin Franklin to investigate. The resulting report debunked Mesmer's practice. By the mid-eighteenth century, British surgeon James Braid reconceptualized mesmerism as a treatment he called hypnosis, named for the Greek god of sleep, Hypnos, reflecting the practice of making leading suggestions to patients in a drowsy state.

When used in psychotherapy, clinical hypnosis has clear benefits. It improves outcomes in everything from pain management to anxiety, depression, insomnia, quitting smoking, and much more. Hypnosis doesn't work for everyone, but depending on a patient's suggestibility and the practitioner's skill, suggestion can have empowering and even pain-alleviating effects. Suggestion and hypnosis are now a regular part of some modern-day mental health care routines and can have measurable effects on brain activity. Research has shown that hypnosis changes activity in a network of brain regions responsible for controlling and monitoring our performance. It narrows our awareness to the task at hand, making us less likely to reflect and thus resist with doubt. We don't question the change or the suggestion, our overthinking slows, and we accept the suggestion on its face.

Hypnosis may also have effects on vision. In 1972, Herschel Leibowitz, a professor of psychology at Pennsylvania State University, had participants with nearsightedness and who were susceptible to hypnosis sit in a dark room where their eyesight was tested with an eye chart. After that, they were hypnotized, then given the suggestion that relaxing would help them see better. After the hypnosis, they were tested again. Their vision seemed to improve slightly, so that they could distinguish objects just slightly better, seeing with 20/60 vision on average after hypnosis, as compared with 20/100 vision before. Leibowitz and his colleague attributed this improvement not to any changes in the shape of the eye or its refractive power but to "neural mechanisms" that process vision, presumably in the brain. However, the participants were allowed to view the same eye chart multiple times, so part of their ability to read it better after hypnosis might have come from memorization, as they had been exposed to it before.

In 1982, Eugene Sheehan and colleagues in the Department of Psychology at Trinity College Dublin, Ireland, did a similar study in those with nearsightedness. One group listened to fifteen minutes of audio suggesting that their vision would improve, while the control group listened to classical music for fifteen minutes. The group that listened to the suggestion audio showed a slight increase in their visual acuity, equivalent to about one line further down the Snellen eye chart, and this was significantly better than those who listened to the classical music. Also, the visual cues they had to identify were shown one at a time to minimize the possibility of memorizing an eye chart. Overall, improvements in such studies are small, with potential alternative explanations including memorization of the eye chart through repeated exposure, or faulty statistical analysis. Frustratingly, studies of how hypnosis may improve hearing have looked for improvements but found none.

Still there is good reason to think that vision can improve, at least with practice, through a process called *perceptual learning.* Aaron Seitz is a professor in the Center for Cognitive and Brain Health at Northeastern University and a former classmate of mine from graduate school days at Boston University. Seitz designed an experiment in which he divided a baseball team into two groups. Half the players were trained for twenty-five minutes per day for a month to recognize subtle visual signals on a screen. This would not be expected to cause any changes in their eyes or their ability to focus on what they were seeing. Instead, the intention was to train their brains to distinguish between objects more effectively, making better use of the same information that was coming from their eyes. The other half of the players were not given the training. At the end of the thirty-day training, the trained players were able to read text 30 to 50 percent farther away than before training. The visual cortex in their brains, which is the part of the brain that processes vision, improved in its ability to process the limited signals provided by the eyes. For several subjects, their post-intervention vision was even better than normal, at 20/7.5 Snellen acuity. That means they could see from 20 feet away what someone with normal vision could see from no farther than 7.5 feet away. *Their brains had literally learned to see better.*

The results we found in Mozambique and Brazil showed that healing prayer could indeed improve vision and hearing, but how it worked remained something of a mystery. Hypnosis and suggestion might account for some of the effects but were unlikely to account for all the improvements. Although in one sense these improvements looked like miracles, we could not test them against the Lambertini criteria, because we had no medical records to identify the nature of the impairments. Indeed, we couldn't rule out conversion disorder being resolved, although we considered that unlikely—the healing prayers were brief, focused on physical conditions, and not addressed to healing psychological trauma that could have led to conversion disorders. We also couldn't rule out the possibility that people were purposely or unintentionally holding back during initial tests before prayer. There was no separate control group, so we couldn't isolate exactly what about the healing prayers seemed to lead to improvements, or whether subjects might have improved apart from the prayers. Candy discussed these points, both the strengths and limitations of the findings, more fully in her book *Testing Prayer* and in her article in the *Southern Medical Journal*.

The question that remains now is, how far can the effects of religious belief and related practices of healing prayer go? In the next chapter, we explore the outer limits of what healing experiences are possible with prayer, looking for cases that pose the biggest challenge for explanation—cases that meet the Lambertini criteria.

10

Miracles Happen

By early 2011, I was still alive and healthy, with no tumor-related symptoms. How that state of affairs came to be, and what it might mean, is a topic I return to later in the next chapter. Meanwhile, Candy had spent seven years collecting data around the world on the effects of prayer for healing. She had just finished writing her book, *Testing Prayer*, in which she described several remarkable cases of healing—and included medical records from before and after prayer. One day, without warning, Candy received an email from a retired Harvard Medical School professor, Martin Moore-Ede. He disclosed that he was one of the anonymous reviewers selected by Harvard University Press to assess whether her book manuscript merited publication. After submitting his review, in which he enthusiastically recommended publication, he gave the press permission to make his identity public. He then wrote directly to Candy. She'd described compelling cases, he noted with admiration. And then he planted a seed that would lodge itself in my brain and my life, shifting my world once again:

"Would you be interested in establishing a non-profit research institute to investigate more cases like these?"

Candy and I discussed the possibility and decided that we'd love to explore it. I was planning to be in Boston that April for an academic conference, so Martin and I met for breakfast at the Sheraton Commander in Cambridge. Two months later, Martin, his wife, Donna, and Brenda Jones, a nursing professor, flew to our house for a meeting. The Global Medical Research Institute (GMRI) was born. By June 2012, we had our non-profit status.

The research goals of GMRI were twofold. First, we wanted to examine cases of healing through prayer where we have thorough medical records, since the lack of medical documentation had limited our understanding of healings we'd observed in places like Cuba and Mozambique. We were interested especially in cases that would meet the Lambertini criteria. The resulting case reports, which we've since published in peer-reviewed medical journals, cannot "prove" scientifically that miracles result from prayer. Individual cases of remarkable healing after prayer are not repeatable events. We cannot—at least ethically—set up an experiment to re-create the conditions and test whether the same individual will again experience healing. To illustrate with a somewhat far-fetched example, if someone claimed to experience miraculous healing of a broken arm, we could theoretically re-create the conditions by intentionally re-breaking their arm—but that's obviously out of the question. And, even in this example, the conditions would not actually be the same, since we'd be comparing prayer for healing of a once-broken and a twice-broken arm.

We needed to take a more interdisciplinary approach that learns from historical and legal standards for evaluating evidence. We asked patients about their experiences during and after prayer, collected medical records from before and after they reported healing, studied differential diagnoses and prognoses, consulted with medical specialists, and looked for other cases of healing from similar conditions. Like the Catholic Church committees that investigate miracle claims, we tried to find ways to explain healings in terms of natural processes and thus *dis*prove the necessity of a miraculous explanation. We then asked: Can we prove, beyond a reasonable doubt, that a miracle occurred in a way that meets the Lambertini criteria? What can we learn from such miracles about how much and what types of healing may be possible?

We then wanted to be able to run randomized controlled clinical trials, the gold standard of scientific research, on the effects of prayer on health outcomes. For clinical trials, I envisioned the kind of measurements we did in Mozambique and Brazil, adding review of medical records, comparison with a control group, and randomization. I was specifically interested in whether people within certain demographics or with certain religious beliefs and/or with certain conditions benefited most from healing prayer. I also envisioned doing more trials on how, under what conditions, and whether healing prayer improved people's overall sense of well-being.

Clinical trials are particularly expensive and complicated. Because we had very little funding, we started GMRI's work with individual case studies. We have now begun to run clinical trials, like the randomized controlled trial of prayer for pain and anxiety in Baltimore, discussed in chapter 6. Most of our effort has, however, focused on investigating individual healings. Anyone who tunes in to a few Christian TV broadcasts or browses church newsletters can easily find many thousands of miracle testimonies. We thought it would be easy to investigate the most compelling claims. We set up a website for GMRI, creating a link where people could describe their healing and upload copies of their medical records. We sat back and waited, nervous that we might soon be overwhelmed by a flood of responses. As it turns out, people are much more eager to share testimonies of miraculous healing than to provide medical records, even with assurances of anonymity. There are a few reasons for this. Many who recover from an illness are understandably eager to get on with their lives. Some fear that pointing to medical evidence suggests a lack of faith. Others feel overwhelmed by the prospect of requesting and deidentifying medical records.

Since miracles did not chase us, we hired research assistants to search for remarkable healings, obtain informed consent to investigate, and track down medical records. It could take hundreds of hours to follow a single lead. After we consulted medical specialists, many seemingly inexplicable cases turned out to have a relatively plausible natural explanation or to be missing information needed to rule out non-miraculous alternatives. That is, until we came across Marsha's case.

In 1959, at age eighteen, Marsha lost vision in both of her eyes over the course of three months. A doctor examined the backs of her eyes and found substantial damage consistent with juvenile macular degeneration. She was declared legally blind, with no hope of recovery. Her parents enrolled her in a school for the blind. There she learned to walk with a cane and read braille. She eventually fell in love, married, and had a child, though she had never seen her husband or daughter. Her husband was a pastor, and they often prayed for Marsha's healing, though seemingly without effect. They even attended meetings of a well-known healing evangelist, again apparently without effect. Another physician reconfirmed the diagnosis in 1971, again noting extensive degeneration of the backs of the eyes. Medically speaking, it is impossible for the eyes to recover from this kind of damage.

One night in 1972, shortly after midnight, as they were kneeling together by the bed to pray before retiring for the night, Marsha's husband prayed for healing of her eyesight as they had done many times before: "Oh, God! You can restore Marsha's eyesight tonight, Lord. I know you can do it! And I pray you will do it tonight."

At the close of the prayer, Marsha opened her eyes and saw her husband kneeling in front of her. This was her first clear visual perception after almost thirteen years of blindness. She gasped.

"I can see! I can see!"

They were both shocked. As Marsha later told us when we interviewed her, "The only healings we knew about were in the Bible." Marsha further described her experience: "What people need to understand is I was blind, totally blind and attended the School for the Blind. I read Braille and walked with a white cane. Never had I seen my husband or daughter's face. I was blind when my husband prayed for me—then just like that—in a moment, after years of darkness I could see perfectly! It was miraculous! My daughter's picture was on the dresser. I could see what my little girl and husband looked like, I could see the floor, the steps. Within seconds, my life had drastically changed. I could see, I could see!"

GMRI obtained copies of Marsha's medical records, with her express informed consent, and consulted with a professor of ophthalmology to understand the case from a medical perspective.

The records show that for at least ten years, there was extensive damage to the back of Marsha's eyes, where light is converted into nerve signals and sent to the brain. If we wondered whether it was a case of psychosomatic illness, specifically a conversion disorder, the records of extensive damage to both of her eyes ruled that out unequivocally. Marsha was blind because her eyes were ruined by the disease, not simply because of emotional distress. We also obtained more recent medical images of the back of Marsha's eyes, after her healing, which show a small amount of scarring where previously there had been extensive damage. Her eyes had healed, but that was impossible—eye damage like that simply cannot be healed, even with the latest medical treatments. We considered every possible natural explanation we could collectively think of. In the end, it was clear that the back of the eyes had been severely diseased, that this would cause blindness, that this fact had been observed multiple times, and that medically there was no hope that she would ever see again. Nevertheless, after prayer for healing, her vision was completely, instantly, and permanently restored.

When we published a detailed report of her case in 2021, Marsha's eyesight had remained intact for forty-nine years. The most likely cause of her blindness was Stargardt disease, a specific genetic disease under the umbrella diagnosis of juvenile macular degeneration. Stargardt disease to this day has no cure and affords no medical hope of recovery.

As inspiring as Marsha's story is, it is not unique. GMRI has now researched a number of comparably compelling cases, which, like Marsha's, we have published in peer-reviewed medical journals along with the corresponding medical records.

In 1995, a boy named Derek was born full-term and apparently healthy, but within a few days he was projectile vomiting. The doctors found that food was not passing through his stomach normally, so when Derek was just six weeks old, he underwent two surgeries. The first was intended to make it easier for food to pass from the stomach to the small intestine, and the second was to prevent reflux. Neither operation resolved the problem.

Eventually the doctors realized that Derek's stomach was paralyzed, a condition called gastroparesis. Normally the stomach contracts to pump food from the stomach to the small intestine, but Derek's stomach wouldn't contract. Any food he swallowed simply sat in the stomach and rotted, until he eventually vomited it. The doctors tried various medications to help the stomach contract, and still nothing helped. The only way Derek could get nutrition was by a tube that went through his nose, passed through his stomach, and delivered liquid food directly into his small intestine. This was not sustainable. The following month, surgeons placed a permanent tube directly into Derek's stomach through his abdominal wall, and then another tube directly into his intestines, again through his abdominal wall. Derek lived the first sixteen years of his life in this state, vacuuming any food he swallowed out of his stomach through one tube, and delivering food to his small intestine through the other tube. He gained weight and grew relatively normally but still couldn't eat normally, and life was complicated by the two tubes protruding from his abdomen.

One day when Derek was sixteen, he went to a prayer meeting. The speaker at that meeting, Robert, told his own life story of how he had been working under a semitruck some years ago, when it fell on his midsection and crushed his intestines. There was only one inch of space between the bottom of the truck and the pavement, and Robert's entire midsection was crushed into that one inch. Robert told how he had been rushed to surgery where doctors removed most of his intestines to save his life, because the intestines that had been crushed were injured beyond repair. Robert survived, but with most of his intestines gone, he couldn't absorb nutrients. He was slowly wasting away. One day, Robert had gone to a prayer meeting, and while he was receiving prayer, he had felt something like a hose uncoiling inside his stomach. After that experience, he suddenly started gaining weight. He was able to absorb nutrients.

Derek listened to Robert tell his story of being crushed by a truck and then healed, and Derek was inspired. The medical journal article GMRI published describes what happened next:

> After the sermon, the evangelist [Robert] talked to the boy and they "compared battle-scars," as both went through several surgeries, developing an instant comradery. He [Robert] asked the

> whole family to gather, and he led a time of PIP [proximal intercessory prayer] (laying hands upon the boy's shoulders). The patient doesn't recall how long the PIP intervention took but mentions that he was prayed for only once. The intercessor prayed that, in the name of Jesus, the boy's stomach be healed. He commanded the healing in the authority and power of Jesus. He made a point of indicating that he had no power or authority to heal, but only with the authority of Jesus Christ, he could command the healing. Halfway through the prayer the boy recalls a shock starting from his right shoulder going down in a diagonal angle across his abdomen and described it as a pulsating and electrical sensation. It surprised the boy, and he reports that he also experienced some pain at the time of the shock. Despite the discomfort, they continued to pray a while longer . . . That night after prayer, he ate a meal for the first time without any complications.

From that day, Derek was able to eat normally and no longer required the tubes. After three months, doctors removed the two tubes from his stomach permanently. In the following seven years up to when the report was written, Derek never had another stomach issue. GMRI obtained copies of Derek's medical records, with his full informed consent, and we asked multiple professors who teach gastroenterology at a medical school to examine the case. They found the case compelling, saying they had not seen one like it, and medically speaking there was no explanation for it. The diagnosis of gastroparesis had been established and confirmed repeatedly. After a few years with the condition, there was no further medical hope of recovery, and Derek and his parents simply had to live with it. But they didn't give up believing that God could heal him. They had no way of knowing that after sixteen years Derek would be instantly, permanently, and completely healed despite having no medical hope.

And then, there was Peter. Peter was born in 2000, seemingly healthy. He was breastfed, but when he was first given solid food at five months of age, he had severe abdominal pain and failed to gain weight. Doctors tried removing dairy, gluten, and fat in case he was having an allergic reaction, but none of this helped. Doctors then surgically implanted a tube in Peter's stomach, but still the pain continued. Unlike Derek, Peter could not even tolerate having food put directly into his intestines. Finally, doctors began injecting nutrition directly into his veins. At sixteen months of age, Peter developed blood problems. His body couldn't make enough of various blood cells. He became anemic and began to lose strength despite intravenous nutrition, which should have been enough. The nerve cells that connected to and activated his muscles became diseased. By age four, he was put in a wheelchair because he didn't have the strength to walk. The following year he lost his fine motor skills and started having seizures. His muscles continued to deteriorate. It was unclear what was causing all these symptoms.

Doctors found an important clue to Peter's diagnosis. Whenever he tried to eat, he had diarrhea with excess fat in it. This meant his digestive system was unable to absorb fats. With this, the doctors suspected a problem with Peter's mitochondria. These are the parts of each cell that, like power plants, make energy, processing things like sugar and fat and making energy available to cells. Without functioning power plants, the cells can't do what is necessary to keep the body running. The whole body is then unable to produce the various chemicals and tissues that it needs. This could explain Peter's inability to digest food or make normal blood components, his seizures, and his extreme weakness and muscle loss. Doctors narrowed the diagnosis down to several possible genetic diseases. They ran a series of tests but couldn't find any known mutations. This is unsurprising, since not all the mutations that cause such diseases have been identified, and more are being discovered regularly.

Doctors eventually concluded that Peter had a mutation that causes a disease of the mitochondria, with one of the genetic mutations that has not yet been characterized. The prognosis was grim: there is no cure and no effective treatment, and 74 percent of patients die by age ten. As Peter's doctor describes: "[Peter] is still under investigation to find an underlying cause. It is still highly likely that by finding a definitive diagnosis, one will not be able to significantly improve his level of function. There is no causal therapy for mitochondriopathy. It is therefore certain/highly likely that [Peter's] level of function will continue to be abnormal or further deteriorate. Finding a diagnosis that can be treated so that [Peter] in any way recovers is completely out of the question."

When Peter was eleven years old, doctors advised his family to make whatever family memories they could while there was still time, as Peter was not expected to live much longer. The family traveled to a healing prayer conference at a large church, with Peter in a wheelchair. GMRI's medical journal article describes what happened next:

> While attending the Healing Rooms at the church, PIP [proximal intercessory prayer] was extended to participating members ("Does anyone need a miracle?"), to which the patient responded by raising his hand. Healing rooms at the church are a multi-faceted experience characterized by worship, dance, painting, prayer, and teaching. The gathering is focused on encounters with God and healing. When questioned by a man about the presenting need for PIP, the patient replied simply, "I can't eat." The intercessor prayed to the Judeo-Christian God: "I pray for new life in [Peter's] stomach and digestive system." No unusual sensations were experienced during PIP. Half an hour following PIP, the patient tolerated oral intake (breadstick) without complications for the first time. Prior to this incident, even small amounts of watermelon would provoke abdominal pain and loose stool.

From that day forward, Peter was able to eat normally and without pain. With Peter's full informed consent, GMRI obtained copies of his medical records from birth and extending ten years after his healing experience. His intravenous nutrition implant was removed three months after his healing because he no longer needed it. The seizures stopped, and his blood numbers normalized within a few months. He gained strength and no longer required a wheelchair. He was well. Peter describes his experience of healing: "The year after my healing was filled with first-time experiences: first day at school, learning how to ride a bike at age 12, ice skating, skiing, sleepovers at a friend's house, handwriting, eating birthday cake, jumping on a trampoline, wilderness camping (away from hospitals!). As a kid, I used to say, 'If I grow up, I'd like to . . .' When I was healed, I could plan for my future: studies, job, relationships. I've completed studies, worked full-time, and traveled worldwide. I'm so grateful."

Peter's case, as well as Marsha's and Derek's, might appropriately be classified as miracles because they meet the standards established by the Lambertini criteria.

Miracles like the stories above are often offered as proof of a religious claim, such as that God exists, or that an exemplary person should be canonized as a saint. Although I am a Christian and believe in God, I affirm that I am not a theologian, and I have no stake in canonization proceedings. As we have seen, there are many cases that have been reported as miracles, but on closer inspection, may have a natural explanation. That does not diminish the joy and relief of those who experience healing, sometimes unexpectedly. Nor does it *dis*prove that God may have worked *through* natural or medical means, or even by intervening supernaturally, to heal.

What the above cases illustrate by meeting the Lambertini criteria is that *we may need to update our beliefs that miracles don't happen*. More healing is possible than we know or realize, and in-person healing prayer practices may have more positive effects than absolute naturalists would feel comfortable acknowledging. No matter how dire our situation, there is still room for hope, and hope itself may improve our health. Our beliefs—including religious or spiritual beliefs that inspire us to pray for healing—affect our brains, and our brains affect our mental and physical health, our sense of well-being, and perhaps even our life expectancy. Some cases defy explanation. Beyond a reasonable doubt, we may reach a verdict that they are miracles. Perhaps science will one day uncover an explanation for this. Maybe it won't. Whatever the case, when our healing comes and cannot be explained medically, that may still be preferable to an absence of healing that we can explain.

GMRI today continues to investigate cases of claimed miraculous healing. While I am a Christian, there are others like Jacalyn Duffin who identify as atheists but nevertheless acknowledge that miracles do happen and find value in investigating them. This is as it should be. A careful investigation of unexplained healing is the domain of scientists and doctors, regardless of religious faith or absence thereof. The mission of GMRI from its founding is to study Christian healing practices. We welcome those who would join our efforts to study such healing practices with curiosity, critical thinking, and an open mind to follow the data wherever they may lead.

11

Believing in Miracles

Following my brain tumor diagnosis, I spent many months and close to a year's salary traveling the world, looking first for my own healing and then trying to understand how beliefs heal. As I write this, over twenty-two years have elapsed since I was diagnosed with a brain tumor. I have now been symptom free since my last seizure on New Year's Day 2004, over twenty-one years ago. Because surgery, radiation, and chemotherapy would not have been curative and carried substantial risks, my doctors and I initially decided to continue monitoring the tumor and consider invasive treatments later as needed to alleviate symptoms. Every three months for the first year after my diagnosis, I continued to undergo MRIs.

In the meantime, I visited healing prayer meetings wherever I heard reports of miracles. I traveled around North America (Washington, Oregon, Illinois, Missouri, British Columbia, South Carolina, Texas, California) and later around the world (Cuba, Brazil, Colombia, Honduras, Nicaragua, Mexico, the United Kingdom, Mozambique, Uganda). Everywhere I went,

I both observed healing prayers and was grateful to receive them. The unusual sensations I experienced while being prayed for were sometimes quite strong, leaving me feeling as if electricity was surging through my body, and this piqued my scientific curiosity as to what was happening, medically and scientifically.

When I was first diagnosed, a friend of mine who was a doctor at the hospital treating me asked the chief of radiology for a second opinion. The mass was not subtle, the chief radiologist told us, and given the seizures I had started having, this was obviously a tumor. But there was still a slim hope—a small chance that this could be a focal, or localized, infection, he told us; if so, it would clear up within three months. But three months later, in November 2003, the mass was still there.

By the end of the first year post-diagnosis, the MRI reading noted that the tumor was not growing. "No interval change seen in the mass within the left medial temporal lobe," the report stated. It continued: "The differential diagnosis for this lesion includes cortical dysplasia and a low grade neoplasm [i.e., glioma]." The cortical dysplasia possibility was new, and I discuss that below. After the first year, I had MRIs every six months, eventually stretching to every year, then every eighteen months. The mass remained visible on MRIs but did not grow.

On September 14, 2010, even as I continued to run MRI scans on hundreds of people for my research, it was my turn to go in for another MRI of my brain. In September 2007, my neurologist had planned two more MRI scans, each at eighteen-month intervals: one in March 2009 and, assuming that scan showed nothing new, the current one.

It was now seven years after my diagnosis, and this was to be my last MRI, though I did not know it at the time. As usual, I picked up the records from the radiologist's office directly. The MRI reading from September 14, 2010, says: "The left hippocampus is slightly smaller than the right, but this may be within normal variation and is stable. No other significant abnormalities of signal intensity are seen . . . Left amygdala and uncus hyperintensity seen on the thin-section T2 and FLAIR images, which is stable. This may be due to mesial temporal sclerosis. No definite involvement of the hippocampi are seen."

The spot that doctors had been calling a tumor for seven years now looked like "mesial temporal sclerosis," or localized scar tissue, nothing more. On the morning of September 23, I met with my neurologist to discuss the MRI results.

I wasn't entirely shocked when I heard what he had to say. He told me that because I had no symptoms and hadn't displayed any for a while, and because there seemed to be only scar tissue where the tumor was, there was no reason to keep doing regular scans. I should call him if I had any new symptoms. Otherwise, he didn't need to see me again, ever. After seven years living under a death sentence, I could simply go out and . . . live. Filled with joy, I went home straightaway to tell Candy.

The tumor was gone, and only scar tissue remained visible on the MRI images, despite the fact that I had no surgery, radiation, or chemotherapy. What had happened to the tumor? I was also curious, supposing the prayers had some effect on my brain, what that effect was, in terms of the biochemistry of neurons. Neuroscientist Dr. David Linden's poignant exploration of his own cancer battle in *The New York Times* inspired me, given that he had survived his cancer for longer than expected. Linden observed that some neurons make direct connections with tumors. When those neurons are silenced, the tumors don't grow as quickly. In other words, the brain can directly influence cancer growth.

I searched the medical literature for any additional clues to what might have happened in my case. I investigated whether the medicine I was prescribed to control seizures, levetiracetam, might have helped shrink the tumor. This is not likely, as levetiracetam doesn't slow tumor growth in patients with more severe glioblastoma tumors, and it doesn't otherwise have properties in common with drugs that treat tumors. Beyond the levetiracetam, I had no conventional or alternative medical treatment—no herbs, no supplements other than an ordinary multivitamin, and no special diet.

Since the only "treatment" I received was prayer, I looked for neuroscientific explanations for how prayer might have stopped the tumor. It turns out that activity in the brain can affect cancer growth, at least in limited ways. Pain-sensing nerve cells in particular make direct connections to some cancer cells and can increase their growth. Various nerve cells also make direct connections to glioma tumor cells, which in turn depend partly on neural signals to grow and spread. That means that blocking some of this neural activity, including with drugs such as perampanel, can slow tumor growth, although such drugs may have severe side effects. Beyond changing the activity of neurons, prayer might have helped relieve my stress, but for the first year after my diagnosis, my stress level was constantly about the highest in my entire life before or since, so that would not have worked in my favor. Still, the prayers often left me with a sense of profound peace. I cannot say for sure, but that reduction in stress might have helped my cortisol levels and strengthened my immune system to carry out immunosurveillance to fight the tumor.

The September 2010 MRI reading from the radiologist began by explaining that I had "epilepsy" of some sort, without any particulars, as the reason for why the MRI was ordered. Taken at face value, the radiology reports indicate that a tumor was present in August 2003, and then, by the time of the last MRI, seven years later, it was a scar. As a scientist, my training is to ask critical questions. Was it really a tumor to begin with? Because I didn't have a biopsy at the time, there is no way to say with 100 percent certainty what exactly it was. There are four possibilities of what could have happened, based on scenarios raised at various points by neurosurgeons, neurologists, and radiologists:

Glioma. This was the original diagnosis, on the basis of a first seizure at age thirty and the MRI findings, which together indicated a tumor. The tumor was classified as low grade because there was no evidence of greater-than-normal blood flow to it, but a low-grade tumor is no less deadly than a high-grade tumor. Despite the "low-grade" technical designation, these tumors eventually transform into high-grade, malignant tumors, invading the surrounding tissue and making complete removal all but impossible. They cause death regardless of treatment, and by the time even low-grade tumors are discovered, they have already begun infiltrating the rest of the brain, and it is too late for a cure. The tumor could perhaps have gone into spontaneous remission, specifically if I had been under sixteen years of age, but I was much older and could find no precedent for spontaneous remission of a glioma in an older person.

Localized infection. Localized infections can cause seizures and be visible on an MRI, but as the head of radiology at Barnes-Jewish Hospital told us, a localized infection would have cleared up by the time of the second MRI, which didn't happen. I had no other indications of a brain infection either at the time or prior to the first seizure.

Mesial temporal sclerosis. This is a kind of scarring that occurs where there's been some previous damage to the brain. Mesial temporal sclerosis can itself cause seizures, but it is not a tumor and is not life-threatening apart from the seizures it causes. The studies I found of mesial temporal sclerosis showed no cases in which the seizures associated with this kind of scarring started after age seventeen, unless there was some kind of head trauma, which I didn't have. Because my two seizures had started at age thirty, not before age seventeen, and I had no head trauma, mesial temporal sclerosis didn't seem viable as an explanation for what had been on the films these past seven years. The radiologist who suggested that it was mesial temporal sclerosis in 2010 noticed apparent scar tissue, but it's not clear he knew that the seizures I was having didn't start until I was thirty, which would have made this diagnosis by itself untenable. The other possibility is that the apparent mesial temporal sclerosis was a scar left over from where the tumor had been located before it was healed. Radiologists typically compare the current MRI with the most recent previous scan, without studying a patient's full medical history. I had, moreover, moved from Missouri to

Indiana in 2006, which further disrupted continuity in tracking my medical progress.

Focal cortical dysplasia. This is a congenital defect in which a small part of the brain grows abnormally in a fetus, and it is not cancerous. When a spot appears in the medial temporal lobe, like mine did, it's nine times more likely to be a glioma than a focal cortical dysplasia. In the 10 percent of patients whose mass in the medial temporal lobe was caused by focal cortical dysplasia, seizures typically start by age five, with 98 percent having seizures before age thirty. The overall higher incidence rate of glioma versus focal cortical dysplasia, combined with my lack of seizures until age thirty, put the odds of my having a focal cortical dysplasia at less than one in five hundred. Put another way, the probability that what I had was a glioma rather than a focal cortical dysplasia was 99.8 percent, leaving a .2 percent chance that I had a focal cortical dysplasia.

Did I really experience a miraculous healing of a glioma, or did I simply beat the odds by having a focal cortical dysplasia? Let's suppose that I had undergone a biopsy. Given that I am alive and symptom free twenty-two years later with no surgery, radiation, or chemotherapy, how likely is it that a biopsy at the time would have shown a glioma versus a focal cortical dysplasia? The answer each one of us gives to that question will depend on our experience, beliefs, and the story we find more plausible. For Jacalyn Duffin's medical colleagues, who as absolute naturalists are convinced that an ordinary explanation must always exist, the only possible story is that it was never a life-threatening tumor to begin with. By extension, even the cases that meet the Lambertini criteria for a miracle must still have a natural explanation, even if we don't know what that explanation is. No amount of evidence that a miracle has occurred would be sufficient to persuade, because the possibility that science *cannot* explain a healing is inconceivable.

Those who believe that miracles can and do happen, albeit rarely, might believe a different story. Asked what my hypothetical biopsy would have shown, believers might think it most likely, given the 99.8 percent probability and my unusual sensory experiences while receiving prayer, to have been a glioma. And given that I am alive, believers might conclude that the most likely reason is that I experienced a miracle. The absolute naturalist might respond that improbable events do sometimes happen. But if one accepts that some healings meet the Lambertini criteria for a miracle, who is to say that one improbable story (that I had a focal cortical dysplasia) is always to be preferred as an explanation over another improbable story (that I had a glioma but was somehow healed)?

The availability heuristic is one of the main reasons why different people may come to different conclusions about whether a particular healing is a miracle. The more we see and hear about modern-day miracles, as I did, the more likely we judge miracles to happen in general, and thus the more likely we will be to perceive that a particular healing experience is a miracle. Conversely, those who have not seen or heard credible accounts of miracles will judge miracles to be less likely overall, and thus that any particular healing is also less likely to be a miracle. In both cases, that leads to a vicious, or virtuous, cycle. For those who believe in miracles, cases that are remarkable—

like Corlien, who experienced healing of Parkinson's disease symptoms—may seem like a miracle even though the dopamine transporter tests still showed evidence of disease. After encountering a case like Corlien's and believing it to be a miracle, the case itself is construed as further evidence that miracles happen. Thus, one will find that future cases like Corlien's may seem even more likely to be miracles.

Those who believe miracles are more likely may also experience confirmation bias as a tendency to discount potential natural processes that could explain the healing. Similarly, those who believe miracles are unlikely or impossible may also experience confirmation bias by discounting evidence that defies an explanation via natural processes. Whether we identify more with absolute naturalists or with believers in the possibility of miracles, *we all base our judgments of how likely a miracle is on how easily we can or cannot recall credible claims of miracles.*

At the time I was diagnosed with a glioma, I had never seen a miracle, nor did I expect to. But my experiences since then have caused my beliefs to evolve. Prior to the diagnosis, I carried out my neuroscience research as a methodological naturalist. As I began to see remarkable healings and investigated cases that met the Lambertini criteria, I could no longer deny that miracles happened. They started to seem more common, and my cognitive map changed to reflect what I was observing. When I considered whether a remarkable healing I was investigating could be a miracle, my availability heuristic led me to conclude that in some cases, had more complete medical records been available, those cases might well have met the Lambertini criteria.

If miracles are rare, but non-miraculous explanations in certain instances are even less plausible, it's unreasonable to exclude from the outset the possibility that a miracle occurred.

I am still 100 percent a scientist. But over the course of my experience investigating claims of miracles, my application of methodological naturalism has become more nuanced. In my neuroscience research, I still assume by default that I am studying natural processes and interpret my results through that lens. But I do not absolutely rule out the possibility that miracles sometimes occur and can be studied—nor that natural explanations of apparent miracles might later be found.

In researching claims of miraculous healing, legal reasoning offers a useful analogy. In a criminal trial, a defendant is presumed innocent until proven guilty beyond a reasonable doubt. The standard of evidence for a conviction is appropriately high. In a civil case, however, the standard is a preponderance of evidence to prove that something is more likely than not. Is it more than 50 percent likely that one party is in the right? If so, the judgment will be in favor of that party. Likewise, when considering a miraculous healing claim, the Lambertini criteria are analogous to the standard in criminal law—proving beyond a reasonable doubt. Once we establish that miracles happen on occasion, the standard of evidence in civil law may be applied. When looking at a case of claimed miraculous healing, if the Lambertini criteria are not met, I ask a follow-up question—is it more likely than not that the case was a miracle? Cases that I would think are more likely than not to be a miracle, even though the Lambertini criteria are not met, are ones that I would say *may* be a miracle. These would, in cases where the evidence for a miracle claim is insufficient, correspond to the "remarkable" designation suggested by Dirk Kruijthoff.

As strange as it may seem at first, redefining miracles in terms of their *probability* of meeting the Lambertini criteria opens up a new possibility of belief and its health benefits. It allows us first and foremost to set aside the question of whether miracles prove that God or the supernatural exists. Of course, these questions are important, but to acknowledge that miracles happen is simply to say that *healings sometimes happen in ways we cannot explain.* This belief that miracles happen can be held by all, from atheists like Jacalyn Duffin to the most fervent of religious believers. We do not have to choose between believing in science and believing in miracles. Science continues to advance tremendously and will continue to do so, but the question of whether the scientific enterprise will, or will not, ultimately explain all miracles remains a matter of belief either way. In the meantime, miracles show us a mystery: that more healing is possible than we know.

Viewing healings as more or less likely to be miracles also frees us from the all-or-nothing barrier. We can believe that a particular healing was absolutely 100 percent a miracle if it meets the Lambertini criteria, or probably a miracle if it seems more likely than not given our prior experience of cases that meet the Lambertini criteria, or not likely to be a miracle.

My own experience of healing offers a window into why we believe, or don't believe, that miraculous healings are likely to happen, and how that can affect our health. I cannot claim that my own healing experience met the Lambertini criteria for a miracle, because I didn't have the biopsy that would have proven the glioma conclusively. But if the case of my healing were a civil trial, the preponderance of evidence and my other experiences of cases that do meet the Lambertini criteria would lead me to consider it more likely than not that my healing was a miracle. In the end, different readers will come to different conclusions about my case; my aim here is *not* to persuade us all that my case must have been a miracle.

Ultimately, my aim is to explain how we develop our beliefs about miracles, and to show how it is reasonable to believe that miracles sometimes happen while setting aside deeper questions such as whether God and the supernatural exist. Openness to belief in miracles matters because it may in turn unlock the health benefits of beliefs we explored in previous chapters. Believing that miraculous healing is possible can inspire hope and resilience in the face of stress, as it did for me. Moreover, belief that spiritual practices can lead to healing may be self-fulfilling, via placebo effects and other effects that we don't understand. Because a belief that miraculous healings may occur is reasonable, we don't have to choose between believing what is true and believing what is good for our health. When it comes to the question of whether miracles happen, we can have both.

AFTERWORD

The Mystery of Healing

MAYBE YOU NEED HEALING. IN THE SEARCH FOR IT, YOUR beliefs do matter. As we saw in chapter 3, our beliefs follow from the stories we construct for ourselves about how the world works. These beliefs in turn influence our health in sometimes surprising and powerful ways. As we focus our attention on evidence of miracles and healing through spiritual practices, our brains may actually help us find the healing we need. James Doty, in his book *Mind Magic*, expands on the concept of manifestation and challenges us to think beyond what we know is possible. The universe doesn't care about us, Doty argues, but we can achieve our goals more effectively by affirming our goals intentionally and embedding them in our subconscious. As we do that, our brains will focus attention automatically on the things we need to process and pursue, to help us attain what we need.

If you need healing or a miracle, the first step is to realize that *remarkable healings and even miracles do happen*. Whatever your religious and spiritual beliefs, the accounts I've shared in this book are true, and many of them I have seen or investigated firsthand. If the story you believe is that you have no hope of healing, you may well be right. But if you accept the evidence that healing is possible, no matter how dire the circumstances, you are also right. *Which story will you believe?*

The next step is to retrain your mind to focus on pursuing healing, as I did, with resilience. I spent many hours reading accounts of miracles both ancient and modern, studying how healings worked, and watching others experience remarkable healings, including many more instances beyond the scope of this book. I retrained my brain to pursue miracles. Setbacks do happen, but the grit that facing them develops can help us persevere. Early in my healing journey, when I faced the setback of a disappointing MRI reading, I had a decision to make: whether to give up and make the most of whatever time I had remaining with my young family, or go for broke. I decided that I had no plan B and resolved to continue pursuing healing with everything I had. I would either be healed or die trying.

This is not to say that you will feel positive and/or determined every day. You will also need to weigh the costs and benefits of holding on to hope. Usually, we talk about hope as a good thing, but emotionally it cuts both ways. Holding on to hope with determination may contribute to healing. Yet if you hope for nothing, you can't be disappointed. There is a certain feeling of safety and even peace that comes from relinquishing hope. My advice is to be aware of what you are deciding, whether hope or acceptance, and make the best decisions possible given your situation. It's also important to know that your feelings will fluctuate and that your decision may seem to change from day to day. We all have good days and bad days.

As you pursue healing, it helps to not be easily put off. This was a hard one for me as a scientist. Early in my quest for healing, I attended some healing prayer meetings that had styles and practices so foreign to me that I felt great discomfort participating. I saw people exhibiting strong emotional and physical reactions. I am usually more reserved, but I tried not to let this contrast disturb me. Ultimately, sticking around was the right choice, and I benefited greatly by pushing past my initial discomfort. If you try new spiritual practices, you, too, may find yourself uncomfortable, and that may be okay. Discomfort often accompanies new experiences, but that's not necessarily a bad thing. Change helps us grow.

Finally, looking back on all the cases of remarkable healings and miracles I saw, one thing stands out: None of them happened to an individual alone. It was always in the context of social interaction or shared spiritual practice. Sometimes the healing came during in-person prayer with other people in a home or a doctor's office, sometimes in a healing prayer meeting in a church, sometimes at a dirt lot in a village under lights powered by a generator. There were always other people with compassion who wanted to help suffering people find hope and healing. When I was desperate for both, I surrounded myself with positivity and those who supported me. I shared my journey, especially with those on the same path. I made friends with another family who were pursuing healing, and we went to healing prayer meetings together. Many people told me that my perseverance inspired them, which helped me find meaning in an otherwise painful experience. Likewise, other people's experiences inspired me, as I realized that I had much to learn from them. *If you want to see more healing, seek it in the company of others.*

Ultimately, there is a mystery to how healing works, and why miracles happen for some and not others. I make no pretense to be any more deserving of a miracle than anyone else. I don't know why I survived while my close friend, the best man in my wedding, passed away from colon cancer in his mid-thirties despite our heartfelt prayers. I am grateful for my healing, though I can never be sure of how or why I experienced it. I do know that many people helped me in countless ways as I pursued healing. I have tried to relay the discoveries and principles that I found enormously helpful along the way. If what I've experienced and learned can help others heal, I offer my perspectives with profound humility.

Since I began this journey, I've seen some healings that fit into my scientific training and others that defy it. As a neuroscientist, it is both fascinating and frustrating to find cases of healing I can't explain. The healings I have witnessed often came in surprising and unpredictable ways. They happened to people from all demographics all over the world. They changed my understanding and the questions I ask. My hope is that this book has done the same for you—inspiring, unsettling, and ultimately expanding what you thought was possible. My friend Randy Clark likes to say "there is more" in reference to healing, and I believe it, because I have seen it. Maybe you will too.

ACKNOWLEDGMENTS

"YOU NEED TO TELL YOUR STORY BEFORE SOMEONE ELSE DOES." Caleb Maskell (Princeton University PhD and Associate National Director of Theology and Education for Vineyard USA) spoke those words to me and Candy over lunch at his place on a warm summer day in Denver in 2022, setting in motion a process that ultimately led to this book. A few months later, when Molly Worthen (award-winning historian, professor at UNC Chapel Hill, *New York Times* columnist, and recently a Christian) called to interview me, I somewhat hesitantly agreed but am glad I did. Molly's nuanced column that Christmas Eve in *The New York Times* about miracles and my research and experiences caught the attention of David Doerrer, who became my literary agent. David's insightful guidance brought me to Karen Rinaldi, Senior Vice President and Executive Editor at HarperCollins. Although David and I initially imagined a memoir, Karen envisioned a more broadly significant synthesis of personal, scientific, and spiritual perspectives. Both Karen Rinaldi and Rachel Kambury at HarperCollins provided indispensable direction on where the book might go and advice on how to get there. Dedi Felman spent many hours patiently listening to my storytelling and expertly assisting me in finding the right words and framing. I deeply appreciate how each of these individuals helped me bring together two parts of my world, the scientific and the spiritual, which until then had coexisted mostly separately.

Many of the ideas I have written about gained refinement through conversation with academic colleagues, all leading experts in their fields of neuroscience, cognitive science, psychology, religious studies, and history. Jordan Grafman (Professor at Northwestern University, previously Chief of the Cognitive Neuroscience Section at the National Institute of Neurological Disorders and Stroke) generously shared his experiences pioneering cognitive neuroscience studies of religion. Rob Goldstone (Distinguished Professor and Chancellor's Professor of Psychological and Brain Sciences, and my neighbor, at Indiana University), invigorated my thinking about placebo effects and offered valuable feedback on the manuscript. Todd Braver (Professor at Washington University in St. Louis), was a great postdoctoral adviser to me decades ago and more recently helped me to nuance my discussion of mindfulness. Matt Chafee (Professor of Neuroscience, University of Minnesota Medical School), along with Joel Wong, Peter Finn, Nicholas Port, Aina Puce, and Richard Nance, all professors at Indiana University, deserve special thanks for stimulating my thinking, with special appreciation to Aina Puce and Richard Nance for comments on the manuscript. I also thank Patrick McNamara (Boston University School of Medicine), Richard Gallagher, MD (Columbia University), and Nathan Strachen (University of Wisconsin–Madison) for useful discussions.

While the writing (and any mistakes) is my own, many physicians and other medical professionals contributed to the book. These include Cliff Brooks, OD, who helped design the Mozambique study and provided feedback on the book manuscript; Dirk Kruijthoff, MD, PhD, whose stimulating conversation guided my thinking about categories of remarkable healings and miracles; Karen Garnaas, MD, whose expertise in neurology helped us find the most interesting cases of healing; Thomas Abell, MD, whose gastroenterology expertise helped us critically evaluate some of the miracle claims; Christina Kile, MD, who provided feedback on the endocrinology discussions; and David Levy, MD, and Jonathan Curtis, MD, both neurosurgeons whose expertise informed the description of my brain tumor and its outcome.

Much of the cognitive neuroscience research at my lab at Indiana University would not have been possible without the excellent trainees who joined me over the years and then went on to greater things—Will Alexander (Professor at Florida Atlantic University), Derek Nee (Professor at Florida State University), Sarah Forster (Pittsburgh VA Hospital), Rena Fukunaga (Centers for Disease Control, formerly Office of the US Surgeon General and Harvard McLean Hospital), and Andy Jahn (University of Michigan Medical School) did much of the cognitive neuroscience work in my lab discussed in this book.

Since the founding of Global Medical Research Institute (GMRI) in 2011, many anonymous individuals have supported our research on case reports of claimed miraculous healing and clinical trials of healing prayer. You know who you are and have my deepest appreciation. My research at Washington University in St. Louis and, since 2006, at Indiana University has been generously supported by the John Templeton Foundation, the Office of Naval Research, the Air Force Office of Scientific Research, the Brain and Behavior Foundation (NARSAD), the

National Institutes of Health (especially NIDA and NIMH), the Intelligence Advanced Research Projects Activity (IARPA), and the Central Intelligence Agency (CIA).

It is a deeply fulfilling privilege to serve as director of GMRI. Randy Clark deserves mention as a leader in the Christian community who saw the value of rigorous medical and scientific investigation of spiritual healing, enough to encourage and support the creation of GMRI as an independent research entity. Martin Moore-Ede, MD, PhD (a professor at Harvard Medical School for twenty-three years before becoming the founding director of GMRI), and Donna Moore-Ede, PhD, catalyzed GMRI's formation. Clarissa Romez, MA, has been an invaluable addition to the GMRI board, helping to organize our research processes and critically evaluate cases. GMRI investigations would not be what they are without our other dedicated board members, radiologist David Zaritzky, MD, Professor of Nursing Brenda Jones, PhD, MSN, CNM, FNP-BC, NHDP-BC—who I also thank for helpful comments on the manuscript. Special thanks to our growing army of GMRI researchers in the US and Brazil—Gabbie Cunha, PT; Aline Correia da Silva, MD; Chrisandra Corneliussen; Alexa Chen; Poincyane Assis-Nascimento, PhD; Neysa Marques, PT; Jim Fernander; Katherine (Kate) Jacobson, MD; Jennifer Zipp, DNP; Brenda Case-Cook, RN, DNP; Megan Schrieber; Zachary Keepers; Andrew Y. Kim; Matthew DiNola; Elisha Barrientos; and numerous volunteers who helped make our clinical trial in Baltimore a reality. I thank our GMRI partners in various churches and other religious organizations, who also saw the benefits of rigorous research in spiritual practices—especially Dave and Taff Harvey, and Josh Clark.

The process of developing this book benefited greatly from support of many in the Christian community, especially Lee Strobel (author of *The Case for Miracles*), Craig Keener (professor and author of *Miracles*), Jay Pathak (National Director of Vineyard USA), Justin Brierly (host of the *Unbelievable* podcast), Cameron Bertuzzi (host of the *Capturing Christianity* podcast), Ryan Bethea (host of the *Exorcist Files* podcast), Kathie Lee Gifford (former host of NBC's *Today* show), Elijah Stephens (producer of the movie *Send Proof*), and Jarrod Anderson (CBN producer of a film on miracles). Special thanks go to my partners with the *Miracle* docuseries (Kimberly Clarke, Ben Kasica, Joe Snyder, Joanne Moody, Josh Silverberg, and many others) for helping bring the GMRI cases into public view. I thank my friends Rob and Patrice Reynolds, Ryan and Stacy Pfeiffer, James and Colleen Nesbit, Steve and Linda Braun, and Doug Johnson for support and camaraderie over the years, which helped keep me sane in the vicissitudes of academic life, faith, and a desperate search for healing. I wish with all my heart that my friend Andrew Snekvik had also found healing. I hope that somehow he can see this book and know how much his memory inspires me.

My daughters Katrina and Sarah sacrificed time with me as I traveled in search of miracles and, as they grew in scientific expertise and spiritual insights of their own, helped me to develop my ideas. Katrina's invitation to discuss the work with her classmates at Harvard led to further conversations useful in refining the book. My parents, Bill Brown and Roxane Martinez, and my sister, Rachel, always supported me and showed me the value of both academic pursuits and faith.

Finally, my utmost gratitude extends to my wife and partner in everything, Candy Gunther Brown, PhD. This book is as much about her experiences as mine, and she blazed the trail for many of the themes discussed here in her own academic work. She made it possible for me to spend extended time writing in addition to teaching, running my lab at the university, and directing GMRI. Much of this book came out of countless conversations with her over many years. She read multiple drafts and provided extensive comments and suggestions that helped make a better book.

NOTES

Introduction: A Neuroscientist's Unplanned Search for Healing

xiii *"something spiritual beyond the natural world":* Gregory A. Smith et al., *Decline of Christianity in the US Has Slowed, May Have Leveled Off: Findings from the 2023-24 Religious Landscape Study*, Pew Research Center, February 26, 2025, https://www.pewresearch.org/wp-content/uploads/sites/20/2025/02/PR_2025.02.26_religious-landscape-study_report.pdf, 6.

xiii *21 percent of the population:* Smith et al., *Decline of Christianity*, 224.

xiii *73 percent of US medical doctors:* Robert D. Orr, "Responding to Patient Beliefs in Miracles," *Southern Medical Journal* 100, no. 12 (2007): 1263–67, https://doi.org/ 10.1097/SMJ.0b013e31815a95cb.

xiii *"experienced a physical healing":* Barna Group, "Most Americans Believe in Supernatural Healing," Barna, September 29, 2016, https://www.barna.com/research/americans-believe-supernatural-healing/.

xiii *those numbers are even higher:* Jonathan Evans et al., "Believing in Spirits and Life After Death Is Common Around the World," Pew Research Center, May 6, 2025, https://www.pewresearch.org/religion/2025/05/06/believing-in-spirits-and-life-after-death-is-common-around-the-world/.

Chapter 1: Should Scientists Study Spiritual Beliefs?

6 *A low-grade glioma:* There were three different types of low-grade gliomas the neurologists said were consistent with what they saw on my scans: an astrocytoma, a ganglioglioma, or an oligodendroglioma. Astrocytomas were known at the time to have a ten-year survival rate up to 20 percent, and low-grade astrocytomas had a mean overall survival of 64.8 months; Lawrence Recht, "Gliomas," *UpToDate* (Wolters Kluwer, 2003). Even with surgery, low-grade astrocytoma patients lived on average 71.4 months; Nader Salari et al., "Patients' Survival with Astrocytoma After Treatment: A Systematic Review and Meta-Analysis of Clinical Trial Studies," *Indian Journal of Surgical Oncology* 13, no. 2 (2022): 329–42, https://doi.org/10.1007/s13193-022-01533-7. Gangliogliomas at the time I was diagnosed had a median survival of 90.3 months, and patients were known to survive as long as thirteen years; R. Hakim et al., "Gangliogliomas in Adults," *Cancer* 79, no. 1 (1997): 127–31. Eleven years after I was diagnosed, patients with low-grade gangliogliomas were found to have a ten-year survival up to 88 percent and to have lived up to twenty years without the disease progressing, with 97 percent having surgery to remove the tumor when they were diagnosed; S. Yust-Katz et al., "Clinical and Prognostic Features of Adult Patients with Gangliogliomas," *Neuro-Oncology* 16, no. 3 (2014): 409–13, https://doi.org/10.1093/neuonc/not169. Oligodendrogliomas have a better prognosis, with a median time to progression of five years (range 0.5 to 14.2 without both chemo- and radiation therapy), and a median overall survival of 16.7 years. Immediate versus deferred treatment did not affect disease-free or overall survival; Jon D. Olson, Elyn Riedel, and Lisa M. DeAngelis, "Long-Term Outcome of

Low-Grade Oligodendroglioma and Mixed Glioma," *Neurology* 54, no. 7 (2000): 1442–48, https://doi.org/10.1212/WNL.54.7.1442.

10 *neuroimaging techniques:* Elizabeth A. Phelps et al., "Performance on Indirect Measures of Race Evaluation Predicts Amygdala Activation," *Journal of Cognitive Neuroscience* 12, no. 5 (2000): 729–38, https://doi.org/10.1162/089892900562552.

10 *our beliefs influence everything:* Harold G. Koenig, Dana E. King, and Verna Benner Carson, *Handbook of Religion and Health*, 2nd ed. (Oxford University Press, 2012).

11 *"Neuroscientists have tended"*: Patrick McNamara et al., "Neuroscientists Must Not Be Afraid to Study Religion," *Nature* 631, no. 8019 (2024): 25–27, https://doi.org/10.1038/d41586-024-02153-7.

11 *As far back as 1633:* Ernan McMullin, ed., *The Church and Galileo* (University of Notre Dame Press, 2005), 4.

11 *Charles Darwin's publication:* Charles Darwin, *The Origin of Species: 150th Anniversary Edition*, with Julian Huxley (Penguin Publishing Group, 2003), 163.

11 *"any theory that denies":* Kenneth K. Bailey, "The Enactment of Tennessee's Antievolution Law," *Journal of Southern History* 16, no. 4 (1950): 472–90.

12 nonoverlapping magisteria *(NOMA) thesis:* Stephen Jay Gould, "Nonoverlapping Magisteria," *Natural History* 106, no. 2 (1997): 16–22.

12 *"God delusion":* Richard Dawkins, "When Religion Steps on Science's Turf: The Alleged Separation Between the Two Is Not So Tidy," *Free Inquiry* 18, no. 2 (1998): 18–19; see generally Richard Dawkins, *The God Delusion* (Bantam Press, 2006).

12 *"the real struggle between":* Richard Lewontin, "Billions and Billions of Demons," *New York Review* (1997), https://www.nybooks.com/articles/1997/01/09/billions-and-billions-of-demons/.

13 *"I don't think there is a God":* Michael Shermer, "Why I Am an Atheist," June 23, 2005, https://michaelshermer.com/articles/why-i-am-an-atheist/.

14 *Consider the claim by flat-earthers:* Joe O'Connal and Colleen Hale-Hodgson, "Images of a Curved and Flat Earth Seen from Space Are Misleading," *Canadian Press*, April 10, 2023, https://www.thecanadianpressnews.ca/fact_checking/images-of-a-curved-and-flat-earth-seen-from-space-are-misleading/article_72b3eaf1-1dce-5897-b3d9-c2242506b6aa.html.

14 *"the evidence presented so far"*: Leanne Roberts et al., "Intercessory Prayer for the Alleviation of Ill Health," *Cochrane Database of Systematic Reviews* 2 (2009), art. no. CD000368, https://doi.org./10.1002/14651858.CD000368.pub3: 15.

14 *Disgruntled researchers who wanted:* Charles Piller, *Doctored: Fraud, Arrogance, and Tragedy in the Quest to Cure Alzheimer's* (Atria Books, 2025), 118.

14 *against herpes zoster (shingles):* Emily Tang et al., "Recombinant Zoster Vaccine and the Risk of Dementia," *Vaccine* 46 (2025): 126673, https://doi.org/10.1016/j.vaccine.2024.126673.

15 *a series of contests for* paradigm *dominance:* Thomas S. Kuhn, *The Structure of Scientific Revolutions* (University of Chicago Press, 1962).

15 *"invok[ed] and permit[ted] supernatural causation":* Kitzmiller v. Dover Area School District, 400 F. Supp. 2d 707 (M.D. Pa. 2005).

16 *"abusing science":* Philip Kitcher, *Abusing Science: The Case Against Creationism* (MIT Press, 1998), 10.
16 *seminal studies on meditation:* John Geirland, "Buddha on the Brain," *Wired*, February 1, 2006, https://www.wired.com/2006/02/dalai/.
17 *meditation affected the brain*: "The Science of Meditation," *Mind & Life* podcast, July 22, 2020, https://podcast.mindandlife.org/richie-davidson/.
17 *combined practices based in Theravada:* Ville Husgafvel, "The 'Universal Dharma Foundation' of Mindfulness-Based Stress Reduction: Non-Duality and Mahāyāna Buddhist Influences in the Work of Jon Kabat-Zinn," *Contemporary Buddhism* 19, no. 2 (2018): 275–326, https://doi.org/10.1080/14639947.2018.1572329.
18 *2003 paper:* Richard J. Davidson et al., "Alterations in Brain and Immune Function Produced by Mindfulness Meditation," *Psychosomatic Medicine* 65, no. 4 (July 2003): 564–70, https://doi.org/10.1097/01.PSY.0000077505.67574.E3.
19 *Kabat-Zinn consciously stripped:* Jon Kabat-Zinn, "Some Reflections on the Origins of MBSR, Skillful Means, and the Trouble with Maps," *Contemporary Buddhism* 12, no. 1 (2011): 281–306, https://doi.org/10.1080/14639947.2011.564844.
19 *keynote speech:* Tenzin Gyatso, the Dalai Lama, "Science at the Crossroads," November 12, 2005, His Holiness the 14th Dalai Lama of Tibet, https://www.dalailama.com/messages/buddhism/science-at-the-crossroads.
20 *"to foster a creative dialogue":* "Our Mission," Mind & Life Institute, December 9, 2000, https://web.archive.org/web/20001209120600/http://www.mindandlife.org/mission.html.
20 *Candy Gunther Brown:* Candy Gunther Brown, *Debating Yoga and Mindfulness in Public Schools: Reforming Secular Education or Reestablishing Religion?* (University of North Carolina Press, 2019).
20 *"nominally secular programs":* Candy Gunther Brown, "Can 'Secular' Mindfulness Be Separated from Religion?," in *Handbook of Mindfulness: Culture, Context, and Social Engagement*, ed. Ronald E. Purser, David Forbes, and Adam Burke (Springer, 2016), 75–94, https://doi.org/10.1007/978-3-319-44019-4_6.
20 *"essence of the Buddha's teachings":* Kabat-Zinn, "Some Reflections on the Origins of MBSR."
21 *"lightening our sense of self":* Daniel Goleman and Richard J. Davidson, *Altered Traits: Science Reveals How Meditation Changes Your Mind, Brain, and Body* (Avery/Penguin Random House, 2017), 153.
21 *"Dan [Goleman] was as he":* Goleman and Davidson, *Altered Traits*, 289.
21 *"pleasant states":* Goleman and Davidson, *Altered Traits*, 6.
21 *"altered traits":* Goleman and Davidson, *Altered Traits*, 7.
21 *basic premise of neuroplasticity:* Gottfried Schlaug et al., "Training-Induced Neuroplasticity in Young Children," *Annals of the New York Academy of Sciences* 1169, no. 1 (2009): 205–8, https://doi.org/10.1111/j.1749-6632.2009.04842.x.
22 *The highest transformations occur:* Goleman and Davidson, *Altered Traits*, 290.
22 *"a radical transformation":* Goleman and Davidson, *Altered Traits*, 42.
22 *The wall between studying effects:* Goleman and Davidson, *Altered Traits*, 289:

"Sciences operate within a web of culture bound assumptions that limit our view of what is possible, most powerfully for the behavior sciences. Modern psychology had not known that Eastern systems offer means to transform a person's very being. When we looked through that alternate Eastern lens, we saw fresh possibilities. By now mounting empirical studies confirm our early hunches: sustained mind training alters the brain both structurally and functionally, proof of concept for the neural basis of altered traits that practitioners' texts have described for millennia . . . Contemplative neuroscience, the emerging specialty which supplies the science behind altered traits, has reached maturity."

22 *"McMindfulness" techniques:* Ronald E. Purser, *McMindfulness: How Mindfulness Became the New Capitalist Spirituality* (Repeater, 2019), 129.

22 *"we know of no Buddhism":* Rupert Gethin, "Cosmology and Meditation: From the Aggañña-Sutta to the Mahāyāna," *History of Religions* 36, no. 3 (1997): 183–217; 188.

23 *"new hybrid form of Buddhism":* David L. McMahan, *The Making of Buddhist Modernism* (Oxford University Press, 2008), Oxford Scholarship Online Abstract.

23 *This relatively new field:* Phelps et al., "Performance on Indirect Measures of Race Evaluation."

24 *70 percent of US scientists:* Anne Marie Conlon, "How Religious Scientists Balance Work and Faith," *Nature* 629, no. 8013 (2024): 957–59, https://doi.org/10.1038/d41586-024-01471-0; Elaine Howard Ecklund et al., "Religion Among Scientists in International Context: A New Study of Scientists in Eight Regions," *Socius: Sociological Research for a Dynamic World* 2 (January 2016): 2378023116664353, https://doi.org/10.1177/2378023116664353.

Chapter 2: How Beliefs Affect the Brain and the Brain Affects Health

27 *a* cognitive map*:* Michael H. Connors and Peter W. Halligan, "Revealing the Cognitive Neuroscience of Belief," *Frontiers in Behavioral Neuroscience* 16 (2022): 926742, https://doi.org/10.3389/fnbeh.2022.926742.

27 *"father of American psychology":* John Snarey and Ashley E. Coleman, "James, William," in *Encyclopedia of Child Behavior and Development*, ed. S. Goldstein and J. A. Naglieri (Springer, 2011): 841–44, https://doi.org/10.1007/978-0-387-79061-9_1552.

28 *Cognitive maps encompass:* "Cognitive Maps in Psychology: Definition, Creation, and Applications," *NeuroLaunch: Gray Matter Matters,* January 14, 2025, https://neurolaunch.com/cognitive-maps-psychology-definition/.

28 *Our maps represent:* Edward C. Tolman, "Cognitive Maps in Rats and Men," *Psychological Review* 55, no. 4 (1948): 189–208, https://doi.org/10.1037/h0061626.

33 *neurophysiology of animal brains:* Joshua W. Brown et al., "Relation of Frontal Eye Field Activity to Saccade Initiation During a Countermanding Task," *Experimental Brain Research* 190, no. 2 (2008): 135–51, https://doi.org/10.1007/s00221-008-1455-0.

34 *a genetic variation related:* Julia Wendt et al., "Genetic Influences on the Ac-

quisition and Inhibition of Fear," *International Journal of Psychophysiology* 98, no. 3 (2015): 499–505, https://doi.org/10.1016/j.ijpsycho.2014.10.007.

Chapter 3: How We Form Beliefs

44 *how jurors form beliefs:* Nancy Pennington and Reid Hastie, "The Story Model for Juror Decision Making," in *Inside the Juror: The Psychology of Juror Decision Making*, ed. Reid Hastie (Cambridge University Press, 1994), 192–224.

45 *When the prosecution presents:* Pennington and Hastie, "Story Model," 211.

46 *four factors of persuasion:* Harold D. Lasswell, "The Structure and Function of Communication in Society," in *The Communication of Ideas* (Institute for Religious and Social Studies, 1948), 216.

47 *He has been doing so ever since:* Dimitrios Kapogiannis et al., "Cognitive and Neural Foundations of Religious Belief," *Proceedings of the National Academy of Sciences* 106, no. 12 (2009): 4876–81, https://doi.org/10.1073/pnas.0811717106.

49 *religious beliefs and rituals grow:* Pascal Boyer, "Religion: Bound to Believe?," *Nature* 455, no. 7216 (2008): 1038–39, https://doi.org/10.1038/4551038a.

49 *shared beliefs can help:* Richard Sosis and Candace Alcorta, "Signaling, Solidarity, and the Sacred: The Evolution of Religious Behavior," *Evolutionary Anthropology: Issues, News, and Reviews* 12, no. 6 (2003): 264–74, https://doi.org/10.1002/evan.10120.

50 *Religious communes survive longer:* Richard Sosis, "Religion and Intragroup Cooperation: Preliminary Results of a Comparative Analysis of Utopian Communities," *Cross-Cultural Research* 34, no. 1 (2000): 70–87, https://doi.org/10.1177/106939710003400105.

50 *Another brain imaging study:* Sam Harris et al., "Functional Neuroimaging of Belief, Disbelief, and Uncertainty," *Annals of Neurology* 63, no. 2 (2008): 141–47, https://doi.org/10.1002/ana.21301.

50 *value of our beliefs:* William James, *The Varieties of Religious Experience: A Study in Human Nature* (Longmans, Green, 1902), 15, https://doi.org/10.1037/10004-000.

52 *Whenever our participants experienced:* Derek Evan Nee, Sabine Kastner, and Joshua W. Brown, "Functional Heterogeneity of Conflict, Error, Task-Switching, and Unexpectedness Effects Within Medial Prefrontal Cortex," *NeuroImage* 54, no. 1 (2011): 528–40, https://doi.org/10.1016/j.neuroimage.2010.08.027.

53 *a research group:* Francesca Lecce et al., "Cingulate Neglect in Humans: Disruption of Contralesional Reward Learning in Right Brain Damage," *Cortex* 62 (January 2015): 73–88, https://doi.org/10.1016/j.cortex.2014.08.008.

53 *the Iowa Gambling Task:* A. Bechara, "Deciding Advantageously Before Knowing the Advantageous Strategy," *Science* 275, no. 5304 (1997): 1293–95, https://doi.org/10.1126/science.275.5304.1293.

55 *We used brain imaging:* Joshua W. Brown and Todd S. Braver, "Learned Predictions of Error Likelihood in the Anterior Cingulate Cortex," *Science* 307, no. 5712 (2005): 1118–21, https://doi.org/10.1126/science.1105783.

57 *when dopamine release:* Shaowen Bao, Vincent T. Chan, and Michael M.

Merzenich, "Cortical Remodelling Induced by Activity of Ventral Tegmental Dopamine Neurons," *Nature* 412, no. 6842 (2001): 79–83.

58 *weapons of mass destruction:* "Iraq's Continuing Program for Weapons of Mass Destruction," October 2002, https://www.odni.gov/files/documents/Iraq_NIE_Excerpts_2003.pdf.

60 *statistically more likely:* "Mortality in the United States, 2023," NCHS Data Brief No. 521, December 2024, https://www.cdc.gov/nchs/data/databriefs/db521.pdf.

61 *avoid confirmation biases:* Richards J. Heuer Jr., *Psychology of Intelligence Analysis* (Martino Fine Books, 2019), 108.

62 *narrows down the range:* John R. Platt, "Strong Inference: Certain Systematic Methods of Scientific Thinking May Produce Much More Rapid Progress Than Others," *Science* 146, no. 3642 (1964): 347–53; T. C. Chamberlin, "The Method of Multiple Working Hypotheses," *Science* 148, no. 3671 (1965): 754–59.

63 *practices Buddhism entails:* Robert Wright, *Why Buddhism Is True: The Science and Philosophy of Meditation and Enlightenment* (Simon & Schuster, 2017), 14.

64 *interpreting sickness as divine punishment:* Samuel R. Weber and Kenneth I. Pargament, "The Role of Religion and Spirituality in Mental Health," *Current Opinion in Psychiatry* 27, no. 5 (2014): 358–63, https://doi.org/10.1097/YCO.0000000000000080.

Chapter 4: How Beliefs Affect Mental Health

69 *"Any sufficiently advanced":* Arthur C. Clarke, *Profiles of the Future: An Inquiry into the Limits of the Possible* (Victor Gollancz, 1999), 1.

71 *how prayer can improve:* Holly Adams et al., "The Effects of Prayer on Attention Resource Availability and Attention Bias," *Religion, Brain & Behavior* 7, no. 2 (2017): 117–33, https://doi.org/10.1080/2153599X.2016.1206612.

72 *using cognitive reframing:* Aaron T. Beck, *Cognitive Therapy and the Emotional Disorders* (Penguin, 1979), 213–14.

73 *Cognitive reframing:* Beck, *Cognitive Therapy, 259–61;* David F. Tolin, "Is Cognitive Behavioral Therapy More Effective Than Other Therapies? A Meta-Analytic Review," *Clinical Psychology Review* 32, no. 6 (2010): 710–20, https://doi.org/ 10.1016/j.cpr.2010.05.003.

74 *mindfulness-based variant of CBT:* "Mindfulness-Based Cognitive Therapy," *Psychology Today*, July 20, 2022, https://www.psychologytoday.com/us/therapy-types/mindfulness-based-cognitive-therapy.

74 *randomized people to receive:* Harold G. Koenig et al., "Effects of Religious Versus Standard Cognitive-Behavioral Therapy on Optimism in Persons with Major Depression and Chronic Medical Illness," *Depression and Anxiety* 32, no. 11 (2015): 835–42, https://doi.org/10.1002/da.22398; Harold G. Koenig et al., "Religious vs. Conventional Cognitive Behavioral Therapy for Major Depression in Persons with Chronic Medical Illness: A Pilot Randomized Trial," *Journal of Nervous & Mental Disease* 203, no. 4 (2015): 243–51, https://doi.org/10.1097/NMD.0000000000000273.

76 *recruited adults:* Peter A. Boelens et al., "A Randomized Trial of the Effect of Prayer on Depression and Anxiety," *International Journal of Psychiatry in Medicine* 39, no. 4 (2009): 377–92, https://doi.org/10.2190/PM.39.4.c.

77 *prayers might cause changes in the brain:* Philip R. Baldwin et al., "Neural Correlates of Healing Prayers, Depression and Traumatic Memories: A Preliminary Study," *Complementary Therapies in Medicine* 27 (August 2016): 123–29, https://doi.org/10.1016/j.ctim.2016.07.002.

78 *"political instability":* Michael Inzlicht et al., "The Need to Believe: A Neuroscience Account of Religion as a Motivated Process," *Religion, Brain & Behavior* 1, no. 3 (2011): 192–212, https://doi.org/10.1080/2153599X.2011.647849; Aaron C. Kay et al., "For God (or) Country: The Hydraulic Relation Between Government Instability and Belief in Religious Sources of Control," *Journal of Personality and Social Psychology* 99, no. 5 (2010): 725–39, https://doi.org/10.1037/a0021140.

78 *how beliefs can affect:* Inzlicht et al., "Need to Believe," 203.

79 *The anterior cingulate:* Cameron S. Carter et al., "Anterior Cingulate Cortex, Error Detection, and the Online Monitoring of Performance," *Science* 280 (1998): 747–49.

80 *neurosurgical treatment:* Lauren T. Brown et al., "Dorsal Anterior Cingulotomy and Anterior Capsulotomy for Severe, Refractory Obsessive-Compulsive Disorder: A Systematic Review of Observational Studies," *Journal of Neurosurgery* 124, no. 1 (2016): 77–89, https://doi.org/10.3171/2015.1.JNS14681.

80 *"I started getting this feeling":* Josef Parvizi et al., "The Will to Persevere Induced by Electrical Stimulation of the Human Cingulate Gyrus," *Neuron* 80, no. 6 (2013): 1359–67, https://doi.org/10.1016/j.neuron.2013.10.057.

83 *"Religious people":* Inzlicht et al., "Need to Believe," 203–5.

85 we could predict: Sarah E. Forster, Peter R. Finn, and Joshua W. Brown, "Neural Responses to Negative Outcomes Predict Success in Community-Based Substance Use Treatment," *Addiction* 112, no. 5 (2017): 884–96, https://doi.org/10.1111/add.13734.

85 *cingulate became more sensitive:* Sarah E. Forster, Peter R. Finn, and Joshua W. Brown, "A Preliminary Study of Longitudinal Neuroadaptation Associated with Recovery from Addiction," *Drug and Alcohol Dependence* 168 (2016): 52–60, https://doi.org/10.1016/j.drugalcdep.2016.08.626.

85 *Buddhist meditation has:* Joshua A. Grant et al., "A Non-Elaborative Mental Stance and Decoupling of Executive and Pain-Related Cortices Predicts Low Pain Sensitivity in Zen Meditators," *Pain* 152, no. 1 (2011): 150–56, https://doi.org/10.1016/j.pain.2010.10.006.

86 *Alcoholics Anonymous (AA):* Marc Galanter et al., "An Initial fMRI Study on Neural Correlates of Prayer in Members of Alcoholics Anonymous," *American Journal of Drug and Alcohol Abuse* 43, no. 1 (2017): 44–54, https://doi.org/10.3109/00952990.2016.1141912; Marc Galanter et al., "Young People in Alcoholics Anonymous: The Role of Spiritual Orientation and AA Member Affiliation," *Journal of Addictive Diseases* 31, no. 2 (2012): 173–82, https://doi.org/10.1080/10550887.2012.665693.

Chapter 5: How Beliefs Affect Well-Being

91 *keeping cortisol levels:* Markus Heinrichs et al., "Social Support and Oxytocin Interact to Suppress Cortisol and Subjective Responses to Psychosocial Stress," *Biological Psychiatry* 54, no. 12 (2003): 1389–98, https://doi.org/10.1016/S0006-3223(03)00465-7.

91 *Male dolphins work together:* Virginia Morell, "Rival Teams of Male Dolphins Form the Animal World's Biggest Social Networks, Long-Running Study Finds," *Science*, August 29, 2022, https://www.science.org/content/article/rival-teams-male-dolphins-form-animal-world-s-biggest-social-networks-long-running.

91 *Pied babbler birds care:* Stephanie L. King, "The Evolutionary Roots of Cooperation," *Current Biology* 32, no. 6 (2022): R249–51, https://doi.org/10.1016/j.cub.2022.02.055.

92 *Social contact enhances:* Eric E. Nelson et al., "Social Re-Orientation and Brain Development: An Expanded and Updated View," *Developmental Cognitive Neuroscience* 17 (2016): 118–27, https://doi.org/10.1016/j.dcn.2015.12.008; Jack L. Andrews et al., "Navigating the Social Environment in Adolescence: The Role of Social Brain Development," *Biological Psychiatry* 89, no. 2 (2021): 109–18, https://doi.org/10.1016/j.biopsych.2020.09.012.

92 *oxytocin may in turn:* Linda Handlin et al., "Human Endogenous Oxytocin and Its Neural Correlates Show Adaptive Responses to Social Touch Based on Recent Social Context," *eLife* 12 (May 2023): 1–27, https://doi.org/10.7554/elife.81197.

92 *oxytocin helps reduce:* C. J. Yatawara et al., "The Effect of Oxytocin Nasal Spray on Social Interaction Deficits Observed in Young Children with Autism: A Randomized Clinical Crossover Trial," *Molecular Psychiatry* 21, no. 9 (2016): 1225–31, https://doi.org/10.1038/mp.2015.162; Kerstin Uvnäs-Moberg et al., "The Yin and Yang of the Oxytocin and Stress Systems: Opposites, Yet Interdependent and Intertwined Determinants of Lifelong Health Trajectories," *Frontiers in Endocrinology* 15 (April 2024): 1–16, https://doi.org/10.3389/fendo.2024.1272270.

92 *human flourishing over eighty years:* Liz Mineo, "Harvard Study, Almost 80 Years Old, Has Proved That Embracing Community Helps Us Live Longer, and Be Happier," *Harvard Gazette*, April 11, 2017, https://news.harvard.edu/gazette/story/2017/04/over-nearly-80-years-harvard-study-has-been-showing-how-to-live-a-healthy-and-happy-life/.

92 *social interactions matter:* Robert Waldinger and Marc Schulz, *The Good Life: Lessons from the World's Longest Scientific Study of Happiness; Create a More Meaningful and Satisfying Life* (Simon & Schuster, 2023), 27.

93 *attending religious services regularly:* Harold G. Koenig et al., "Does Religious Attendance Prolong Survival? A Six-Year Follow-Up Study of 3,968 Older Adults," *Journals of Gerontology Series A: Biological Sciences and Medical Sciences* 54, no. 7 (1999): M370–76, https://doi.org/10.1093/gerona/54.7.M370.

93 *People who participate:* Pew Research Center, *Religion's Relationship to Happiness, Civic Engagement and Health Around the World*, January 31, 2019, https://www.pewresearch.org/wp-content/uploads/sites/20/2019/01/Wellbeing-report-1-25-19-FULL-REPORT-FOR-WEB.pdf.

93 *social interactions and the oxytocin release:* Ashwin A. Kotwal et al., "A Peer Intervention Reduces Loneliness and Improves Social Well-Being in Low-Income Older Adults: A Mixed-Methods Study," *Journal of the American Geriatrics Society* 69, no. 12 (2021): 3365–76, https://doi.org/10.1111/jgs.17450.

93 *their brains showed:* Raymond L. Neubauer, "Prayer as an Interpersonal Relationship: A Neuroimaging Study," *Religion, Brain & Behavior* 4, no. 2 (2014): 92–103, https://doi.org/10.1080/2153599X.2013.768288; Uffe Schjoedt et al., "Highly Religious Participants Recruit Areas of Social Cognition in Personal Prayer," *Social Cognitive and Affective Neuroscience* 4, no. 2 (2009): 199–207, https://doi.org/10.1093/scan/nsn050.

96 *respondents were asked:* Daniel Escher, "How Does Religion Promote Forgiveness? Linking Beliefs, Orientations, and Practices," *Journal for the Scientific Study of Religion* 52, no. 1 (2013): 100–19, https://doi.org/10.1111/jssr.12012.

97 *forgiveness activated their brains:* Emiliano Ricciardi et al., "How the Brain Heals Emotional Wounds: The Functional Neuroanatomy of Forgiveness," *Frontiers in Human Neuroscience* 7 (2013), https://doi.org/10.3389/fnhum.2013.00839.

97 *general tendency to forgive:* Haijiang Li and Jiamei Lu, "The Neural Association Between Tendency to Forgive and Spontaneous Brain Activity in Healthy Young Adults," *Frontiers in Human Neuroscience* 11, no. 561 (November 2017): 1–7, https://doi.org/10.3389/fnhum.2017.00561.

97 *forgiveness reduces our anxiety:* Jichan J. Kim et al., "Indirect Effects of Forgiveness on Psychological Health Through Anger and Hope: A Parallel Mediation Analysis," *Journal of Religion and Health* 61, no. 5 (2022): 3729–46, https://doi.org/10.1007/s10943-022-01518-4.

97 *forgiveness is associated:* Yu-Rim Lee and Robert D. Enright, "A Meta-Analysis of the Association Between Forgiveness of Others and Physical Health," *Psychology & Health* 34, no. 5 (2019): 626–43, https://doi.org/10.1080/08870446.2018.1554185; Jennifer P. Friedberg, Sonia Suchday, and Danielle V. Shelov, "The Impact of Forgiveness on Cardiovascular Reactivity and Recovery," *International Journal of Psychophysiology* 65, no. 2 (2007): 87–94, https://doi.org/10.1016/j.ijpsycho.2007.03.006.

97 *"Resentment is like swallowing":* Quote origin: "Resentment Is Like Taking Poison and Waiting for the Other Person to Die," QuoteInvestigator, August 19, 2017, https://quoteinvestigator.com/2017/08/19/resentment/.

98 *by studying gratitude:* Robert A. Emmons and Teresa T. Kneezel, "Giving Thanks: Spiritual and Religious Correlates of Gratitude," *Journal of Psychology and Christianity* 24, no. 2 (2005): 140–48.

98 *effects of gratitude on mental health:* Y. Joel Wong et al., "Does Gratitude Writing Improve the Mental Health of Psychotherapy Clients? Evidence from a Randomized Controlled Trial," *Psychotherapy Research* 28, no. 2 (2018): 192–202, https://doi.org/10.1080/10503307.2016.1169332.

100 *gratitude changes:* T. Sharot et al., "Neural Mechanisms Mediating Optimism Bias," *Nature* 450, no. 7166 (2007): 102–5, https://doi.org/10.1038/nature06280.

100 *It activates when:* Amit Etkin et al., "Resolving Emotional Conflict: A Role for the Rostral Anterior Cingulate Cortex in Modulating Activity in the Amygdala," *Neuron* 51, no. 6 (2006): 871–82, https:// 10.1016/j.neuron.2006.07.029.

101 Grit *is a personality trait:* Angela L. Duckworth et al., "Grit: Perseverance and Passion for Long-Term Goals," *Journal of Personality and Social Psychology* 92, no. 6 (2007): 1087–1101, https://doi.org/10.1037/0022-3514.92.6.1087.

101 *beliefs can provide:* Fábio Duarte Schwalm et al., "Is There a Relationship Between Spirituality/Religiosity and Resilience? A Systematic Review and Meta-Analysis of Observational Studies," *Journal of Health Psychology* 27, no. 5 (2022): 1218–32, https://doi.org/10.1177/1359105320984537.

101 *higher levels of grit:* Katherine A. Traino et al., "The Role of Grit in Health Care Management Skills and Health-Related Quality of Life in College Students with Chronic Medical Conditions," *Journal of Pediatric Nursing* 46 (May 2019): 72–77, https://doi.org/10.1016/j.pedn.2019.02.035.

102 *those with higher levels of grit:* Traino et al., "Role of Grit."

102 *Having a purpose:* Christina M. Sharkey et al., "Grit, Illness-Related Distress, and Psychosocial Outcomes in College Students with a Chronic Medical Condition: A Path Analysis," *Journal of Pediatric Psychology* 43, no. 5 (2018): 552–60, https://doi.org/10.1093/jpepsy/jsx145.

102 *satisfaction comes from engaging:* Martha Kent et al., "Goal-Directed Resilience in Training (GRIT): A Biopsychosocial Model of Self-Regulation, Executive Functions, and Personal Growth (*Eudaimonia*) in Evocative Contexts of PTSD, Obesity, and Chronic Pain," *Behavioral Sciences* 5, no. 2 (2015): 264–304, https://doi.org/10.3390/bs5020264.

102 *Unpredictable chronic mild stress:* Yann S. Mineur et al., "Effects of Unpredictable Chronic Mild Stress on Anxiety and Depression-Like Behavior in Mice," *Behavioural Brain Research* 175, no. 1 (2006): 43–50, https://doi.org/10.1016/j.bbr.2006.07.029.

103 *The resilient mice:* Anand Gururajan et al., "Resilience to Chronic Stress Is Associated with Specific Neurobiological, Neuroendocrine and Immune Responses," *Brain, Behavior, and Immunity* 80 (August 2019): 583–94, https://doi.org/10.1016/j.bbi.2019.05.004.

103 *stress can cause:* Andrew Holmes and Cara L. Wellman, "Stress-Induced Prefrontal Reorganization and Executive Dysfunction in Rodents," *Neuroscience & Biobehavioral Reviews* 33, no. 6 (2009): 773–83, https://doi.org/10.1016/j.neubiorev.2008.11.005.

104 *stressful situation:* Shabnam Hossein et al., "Effects of Acute Stress and Depression on Functional Connectivity Between Prefrontal Cortex and the Amygdala," *Molecular Psychiatry* 28, no. 11 (2023): 4602–12, https://doi.org/10.1038/s41380-023-02056-5.

104 *dampening the amygdala response:* One of the ways that you can predict an individual's likelihood of successfully beating addiction is by looking at how strongly their amygdala reacts when they are under stress. Individuals whose amygdalae respond less strongly to bad things happening are usually better situated to beat addiction. Those whose amygdalae have a strong response to stress often have more trouble abstaining from drugs and alcohol. It's as if addicts are unable to curb their overwhelming emotional response to stress, even if that means that they end up trying to alleviate their pain by reaching for something as unhealthy as the substance they are addicted to; Sarah E. Forster, Peter R. Finn, and Joshua W. Brown, "Neural Responses to Negative Outcomes Predict Success in Community-Based Substance Use Treatment," *Addiction* 112, no. 5 (2017): 884–96, https://doi.org/10.1111/add.13734.

105 *emotionally regulate ourselves:* Kelly M. Moench and Cara L. Wellman, "Differential Dendritic Remodeling in Prelimbic Cortex of Male and Female Rats

During Recovery from Chronic Stress," *Neuroscience* 357 (August 2017): 145–59, https://doi.org/10.1016/j.neuroscience.2017.05.049.

105 We can develop grit: Karen Mueller et al., "An Online Intervention Increases Empathy, Resilience, and Work Engagement Among Physical Therapy Students," *Journal of Allied Health* 47, no. 3 (2018): 196–203; Mae-Hyang Hwang and JeeEun Karin Nam, "Enhancing Grit: Possibility and Intervention Strategies," in *Multidisciplinary Perspectives on Grit*, ed. Llewellyn Ellardus Van Zyl et al. (Springer, 2021), 77–93, https://doi.org/10.1007/978-3-030-57389-8_5; See generally Mary C. Murphy, *Cultures of Growth: How the New Science of Mindset Can Transform Individuals, Teams, and Organizations* (Simon & Schuster, 2024).

108 *Memory recall depends:* David Badre et al., "Dissociable Controlled Retrieval and Generalized Selection Mechanisms in Ventrolateral Prefrontal Cortex," *Neuron* 47, no. 6 (2005): 907–18, https://doi.org/10.1016/j.neuron.2005.07.023.

109 the prayers helped: Suzana Mara Cordeiro Eloia et al., "Religious Coping and Hope in Chronic Kidney Disease: A Randomized Controlled Trial," *Revista da Escola de Enfermagem da USP* 55 (2021): e20200368, https://doi.org/10.1590/1980-220x-reeusp-2020-0368.

110 *learned helplessness:* Martin E. P. Seligman, "Learned Helplessness," *Annual Review of Medicine* 23, no. 1 (1972): 407–12; Jo Nash, "The 5 Founding Fathers and a History of Positive Psychology," February 12, 2015, https://positivepsychology.com/founding-fathers/.

110 *unpleasant experiences:* Donald S. Hiroto and Martin E. Seligman, "Generality of Learned Helplessness in Man," *Journal of Personality and Social Psychology* 31, no. 2 (1975): 311–27, https://doi.org/10.1037/h0076270.

Chapter 6: How Beliefs Affect Physical Health

111 *diagnosed with Hodgkin's lymphoma:* Anita Moorjani, *Dying to Be Me: My Journey from Cancer, to Near Death, to True Healing*, with Wayne W. Dyer (Hay House, 2022).

112 *greater spirituality:* Harold G. Koenig, "Religion, Spirituality, and Health: The Research and Clinical Implications," *ISRN Psychiatry* (December 2012): 1–33, https://doi.org/10.5402/2012/278730.

112 *patient's spiritual and religious history:* Claudia Kalb, "Faith & Healing," *Newsweek*, November 9, 2003, https://www.newsweek.com/faith-healing-133365.

112 *patient with advanced lymphosarcoma:* Bruno Klopfer, "Psychological Variables in Human Cancer," *Journal of Projective Techniques* 21, no. 4 (1957): 331–40, https://doi.org/10.1080/08853126.1957.10380794.

115 *immunosuppression increases:* T. Vial, "Immunosuppressive Drugs and Cancer," *Toxicology* 185, no. 3 (2003): 229–40, https://doi.org/10.1016/S0300-483X(02)00612-1.

115 *deliberately infected the leg:* Ian D. Davis, "An Overview of Cancer Immunotherapy," *Immunology & Cell Biology* 78, no. 3 (2000): 179–95, https://doi.org/10.1046/j.1440-1711.2000.00906.x; Dupré de Lisle, *Traité sur le vice Cancéreux* [Treatise on Cancerous Vice] (Couturier, 1774), 158–59, https://gallica.bnf.fr/ark:/12148/bpt6k1504862g#.

115 *administered inactivated bacterial strains:* Activating the immune system

to fight cancer is a common cause of spontaneous remissions from cancer, though not the only cause. Although rare (one per 60,000 to 100,000 cancer cases), spontaneous remissions do occur. They are, simply by definition, categorized as unexplained because no officially recognized effective treatment was given, even if there is some indication as to the cause. William Bradley Coley in 1891 developed Coley's Toxin, a mix of two inactivated strains of bacteria, which he used to treat a cancer patient successfully. Throughout the twentieth century, subsequent attempts to treat cancer patients by deliberately infecting them with specific bacterial and virus strains continued. There is some evidence that biopsies alone, by damaging cancer cells, can cause them to release their antigens into the surrounding tissue, thus occasionally provoking an immune response and subsequent spontaneous remission; Gudapureddy Radha and Manu Lopus, "The Spontaneous Remission of Cancer: Current Insights and Therapeutic Significance," *Translational Oncology* 14, no. 9 (2021): 101166, https://doi.org/10.1016/j.tranon.2021.101166.

116 *specially modified immune cells:* Steven A. Rosenberg et al., "Use of Tumor-Infiltrating Lymphocytes and Interleukin-2 in the Immunotherapy of Patients with Metastatic Melanoma," *New England Journal of Medicine* 319, no. 25 (1988): 1676–80, https://doi.org/10.1056/NEJM198812223192527.

116 *strengthens the body's own immune system:* Charles Graeber, *The Breakthrough: Immunotherapy and the Race to Cure Cancer* (Twelve, 2018).

116 *checkpoint-ignoring immune cells:* Theresa L. Hunter et al., "In Vivo CAR T Cell Generation to Treat Cancer and Autoimmune Disease," *Science* 388, no. 6753 (2025): 1311–17, https://doi.org/10.1126/science.ads8473.

117 *people with organ transplants:* Claire M. Vajdic and Marina T. Van Leeuwen, "Cancer Incidence and Risk Factors After Solid Organ Transplantation," *International Journal of Cancer* 125, no. 8 (2009): 1747–54, https://doi.org/10.1002/ijc.24439.

117 *number of cancer cells increases:* O. J. Finn, "Immuno-Oncology: Understanding the Function and Dysfunction of the Immune System in Cancer," *Annals of Oncology* 23, supp. 8 (2012): 6–9, https://doi.org/10.1093/annonc/mds256.

117 *role of mental well-being:* Kirk W. Brown et al., "Psychological Distress and Cancer Survival: A Follow-Up 10 Years After Diagnosis," *Psychosomatic Medicine* 65, no. 4 (2003): 636–43, https://doi.org/10.1097/01.PSY.0000077503.96903.A6.

118 *The fewer symptoms of depression:* Brown et al., "Psychological Distress," 639.

118 *depression is associated:* J. Stephen McDaniel et al., "Cancer and Depression: Theory and Treatment," *Psychiatric Annals* 27, no. 5 (1997): 360–64, https://doi.org/10.3928/0048-5713-19970501-12.

118 *especially in older rats:* Robert M. Sapolsky and Thomas M. Donnelly, "Vulnerability to Stress-Induced Tumor Growth Increases with Age in Rats: Role of Glucocorticoids," *Endocrinology* 117, no. 2 (1985): 662–66, https://doi.org/10.1210/endo-117-2-662.

118 *In a population of cancer patients:* Santiago Allende et al., "Evening Salivary Cortisol as a Single Stress Marker in Women with Metastatic Breast Cancer," *Psychoneuroendocrinology* 115 (May 2020): 104648, https://doi.org/10.1016/j.psyneuen.2020.104648.

119 *elevated cortisol levels and the progression of cancer:* Sandra E. Sephton et al., "Diurnal Cortisol Rhythm as a Predictor of Lung Cancer Survival," *Brain, Behavior, and Immunity* 30 (March 2013): S163–70, https://doi.org/10.1016/j.bbi.2012.07.019.

119 *24.3 percent of US adults:* Sherry L. Murphy et al., "Mortality in the United States, 2023," NCHS Data Brief No. 521, December 2024, https://www.cdc.gov/nchs/data/databriefs/db521.pdf; Jacqueline W. Lucas and Inderbir Sohi, "Chronic Pain and High-Impact Chronic Pain in U.S. Adults, 2023," NCHS Data Brief No. 518, November 2024, https://www.cdc.gov/nchs/products/databriefs/db518.html.

122 *medical professionals:* Salvatore Mangione et al., "Out of Touch," *JAMA* 331, no. 9 (2024): 729, https://doi.org/10.1001/jama.2024.0888.

123 proximal intercessory prayer*:* Candy Gunther Brown et al., "Study of the Therapeutic Effects of Proximal Intercessory Prayer (STEPP) on Auditory and Visual Impairments in Rural Mozambique," *Southern Medical Journal* 103, no. 9 (2010): 864–69, https://doi.org/10.1097/SMJ.0b013e3181e73fea.

123 *a prayer- gauge test:* John Tyndall, *The Prayer-Gauge Debate* (1876; HardPress Limited, 2020).

123 *distant intercessory prayer for coronary care patients:* Randolph C. Byrd, "Positive Therapeutic Effects of Intercessory Prayer in a Coronary Care Unit Population," *Southern Medical Journal* 81, no. 7 (1988): 826–29.

123 *study was criticized:* Benedict Carey, "Can Prayers Heal? Critics Say Studies Go Past Science's Reach," *New York Times,* October 10, 2004, https://www.nytimes.com/2004/10/10/health/can-prayers-heal-critics-say-studies-go-past-sciences-reach.html.

124 *believe that prayer would improve symptoms:* Candy Gunther Brown, *Testing Prayer: Science and Healing* (Harvard University Press, 2012), 372.

124 *discouraged further studies:* Leanne Roberts, Irshad Ahmed, and Andrew Davison, "Intercessory Prayer for the Alleviation of Ill Health," *Cochrane Database of Systematic Reviews* 2 (2009), art. no. CD000368, doi:10.1002/14651858.CD000368.pub3: 20.

125 *his 1974 book* Healing*:* Francis MacNutt, *Healing,* Rev. & Expanded (Ave Maria Press, 1999).

125 *effects of prayer on patients:* Dale A. Matthews et al., "Effects of Intercessory Prayer on Patients with Rheumatoid Arthritis," *Southern Medical Journal* 93, no. 12 (2000): 1177–86.

126 *story of a local minister:* "The Life and Conduct of Dr. Rowland Taylor of Hadley," in *Foxe's Book of Martyrs,* by John Foxe (1563), 412–18, https://archive.org/details/foxesbookofmartyrs_201708/page/n411/mode/2up.

127 *how religious beliefs might:* Katja Wiech et al., "An fMRI Study Measuring Analgesia Enhanced by Religion as a Belief System," *Pain* 139, no. 2 (2008): 467–76, https://doi.org/10.1016/j.pain.2008.07.030.

Chapter 7: Placebo Effects

132 the sham surgery patients*:* Raine Sihvonen et al., "Arthroscopic Partial Meniscectomy Versus Sham Surgery for a Degenerative Meniscal Tear," *New England Journal of Medicine* 369, no. 26 (2013): 2515–24, https://doi.org/10.1056/nejmoa1305189.

132 *meniscus surgery leads:* Jianxiong Ma et al., "Medical Exercise Therapy Alone Versus Arthroscopic Partial Meniscectomy Followed by Medical Exercise Therapy for Degenerative Meniscal Tear: A Systematic Review and Meta-Analysis of Randomized Controlled Trials," *Journal of Orthopaedic Surgery and Research* 15, no. 1 (2020), https://doi.org/10.1186/s13018-020-01741-3.

132 *placebo effects are often spoken of dismissively:* A. Hróbjartsson and P. C. Gøtzsche, "Is the Placebo Powerless? An Analysis of Clinical Trials Comparing Placebo with No Treatment," *New England Journal of Medicine*, no. 344 (2001): 1594–602, https://doi.org/10.1056/nejm200105243442106; Wayne B. Jonas, "The Myth of the Placebo Response," *Frontiers in Psychiatry*, no. 10 (2019): 577, https://doi.org/10.3389/fpsyt.2019.00577.

133 *"I will please":* Jeremy Howick, "The Fascinating Story of Placebos—and Why Doctors Should Use Them More Often," *The Conversation*, January 6, 2021, https://theconversation.com/the-fascinating-story-of-placebos-and-why-doctors-should-use-them-more-often-149945.

133 *ordinary hand lotion:* Tor D. Wager et al., "Placebo-Induced Changes in fMRI in the Anticipation and Experience of Pain," *Science* 303, no. 5661 (2004): 1162–67, https://doi.org/10.1126/science.1093065.

134 *where and how pain is represented:* Andrew Jahn et al., "Distinct Regions Within Medial Prefrontal Cortex Process Pain and Cognition," *Journal of Neuroscience* 36, no. 49 (2016): 12385–92, https://doi.org/10.1523/JNEUROSCI.2180-16.2016.

136 *naloxone dramatically weakens:* Naloxone also weakens pain relief by disrupting the way some parts of the brain interact with the brain stem where pain is processed; Falk Eippert et al., "Activation of the Opioidergic Descending Pain Control System Underlies Placebo Analgesia," *Neuron* 63, no. 4 (2009): 533–43, https://doi.org/10.1016/j.neuron.2009.07.014.

136 *Effective placebos increase:* Fabrizio Benedetti and Martina Amanzio, "The Placebo Response: How Words and Rituals Change the Patient's Brain," *Patient Education and Counseling* 84, no. 3 (2011): 413–19, https://doi.org/10.1016/j.pec.2011.04.034.

138 *human body has:* Fabrizio Benedetti et al., "Nonopioid Placebo Analgesia Is Mediated by CB1 Cannabinoid Receptors," *Nature Medicine* 17, no. 10 (2011): 1228–30, https://doi.org/10.1038/nm.2435.

138 *placebo pills given as antidepressants:* Arif Khan and Walter A. Brown, "Antidepressants Versus Placebo in Major Depression: An Overview," *World Psychiatry* 14, no. 3 (2015): 294–300, https://doi.org/10.1002/wps.20241.

139 *placebos can be:* Khan and Brown, "Antidepressants Versus Placebo."

140 *46 percent of those taking the placebo:* Kevin Olden et al., "Patient Satisfaction with Alosetron for the Treatment of Women with Diarrhea-Predominant Irritable Bowel Syndrome," *American Journal of Gastroenterology* 97, no. 12 (2002): 3139–46, https://doi.org/10.1111/j.1572-0241.2002.07111.x.

141 *simply believing:* William G. Ondo et al., "Selegiline Orally Disintegrating Tablets in Patients with Parkinson Disease and 'Wearing Off' Symptoms," *Clinical Neuropharmacology* 30, no. 5 (2007): 295–300, https://doi.org/10.1097/WNF.0b013e3180616570.

141 *Parkinson's disease symptoms improve:* Fabrizio Benedetti et al., "Placebo-Responsive Parkinson Patients Show Decreased Activity in Single Neurons of

Subthalamic Nucleus," *Nature Neuroscience* 7, no. 6 (2004): 587–88, https://doi.org/10.1038/nn1250.

141 *ulcerative colitis and high blood pressure:* Khan and Brown, "Antidepressants Versus Placebo."

141 *placebo effects and cancer:* G. Chvetzoff and I. F. Tannock, "Placebo Effects in Oncology," *JNCI: Journal of the National Cancer Institute* 95, no. 1 (2003): 19–29, https://doi.org/10.1093/jnci/95.1.19.

142 placebo effects can relieve symptoms*:* Cláudia Carvalho et al., "Open-Label Placebo Treatment in Chronic Low Back Pain: A Randomized Controlled Trial," *Pain* 157, no. 12 (2016): 2766–72, https://doi.org/10.1097/j.pain.0000000000000700.

142 *"her doctors":* Alexandra Sifferlin, "People Are Now Taking Placebo Pills to Deal with Their Health Problems—and It's Working," *Time*, August 23, 2018, https://time.com/5375724/placebo-bill-health-problems/.

142 *IBS patients in the same study:* Ted J. Kaptchuk et al., "Placebos Without Deception: A Randomized Controlled Trial in Irritable Bowel Syndrome," *PLoS ONE* 5, no. 12 (2010): e15591, https://doi.org/10.1371/journal.pone.0015591.

Chapter 8: Mind-Body Effects

149 *associated conversion disorder with hysteria:* T. R. J. Nicholson et al., "Conversion Disorder: A Problematic Diagnosis," *Journal of Neurology, Neurosurgery & Psychiatry* 82, no. 11 (2011): 1267–73, https://doi.org/10.1136/jnnp.2008.171306.

150 *a twenty-three-year-old woman:* Abdul Ahad Qazizada, "Sudden Vision Loss: A Case Report and Overview of Conversion Disorder," *Consultant360* 54, no. 8 (2014), https://www.consultant360.com/articles/sudden-vision-loss-case-report-and-overview-conversion-disorder.

151 *For the absolute naturalist:* Piotr Sikorski et al., "The Causes, Diagnostics, and Treatment of Psychogenic Blindness—a Systematic Review," *Current Problems of Psychiatry* 24 (October 2023): 246–52, https://doi.org/10.12923/2353-8627/2023-0023.

154 *Duffin describes her research:* Jacalyn Duffin, *Medical Miracles: Doctors, Saints, and Healing in the Modern World* (Oxford University Press, 2009).

155 *"These events were miracles":* Duffin, *Medical Miracles*, 183.

155 *"utterly pervades the scientific community":* Duffin, *Medical Miracles*, 187.

156 *evaluate evidence:* The reader may note the implicit reference here to Bayesian probability, which suggests that how likely we judge something to be (e.g., a supernatural event) depends on both the strength of the evidence in a particular case and our prior beliefs, in general, about how likely it is that such events can ever happen. If our prior belief is that such events can never happen, then no amount of evidence will be convincing.

156 *what can happen medically:* Hanneke Klein-den Hertog, "Healing After Prayer: Dutch GP Did PhD Research After It," CNE News, June 17, 2023, https://cne.news/article/3213-healing-after-prayer-dutch-gp-did-phd-research-after-it.

156 *83 cases of healing through prayer:* Dirk J. Kruijthoff et al., "Prayer and Healing: A Study of 83 Healing Reports in the Netherlands," *Religions* 13, no. 11 (2022): 1056, https://doi.org/10.3390/rel13111056.

157 *a woman with Parkinson's disease:* Dirk J. Kruijthoff et al., "'My Body Does Not Fit in Your Medical Textbooks': A Physically Turbulent Life with an Unexpected Recovery from Advanced Parkinson Disease After Prayer," *Advances in Mind Body Medicine* 35, no. 2 (2021): 4 –13, PMID: 33620331.

158 *"my body does not fit":* Kruijthoff et al., "'My Body Does Not Fit.'"

158 *seventy-two cases of healing:* "The Miracles of Lourdes," Lourdes Sanctuaire, https://www.lourdes-france.org/en/the-miracles-of-lourdes/.

Chapter 9: Chasing Miracles

162 *effects that might confound the results:* Candy Gunther Brown, *Testing Prayer: Science and Healing* (Harvard University Press, 2012), 194–233; Candy Gunther Brown et al., "Study of the Therapeutic Effects of Proximal Intercessory Prayer (STEPP) on Auditory and Visual Impairments in Rural Mozambique," *Southern Medical Journal* 103, no. 9 (2010): 864–69, https://doi.org/10.1097/SMJ.0b013e3181e73fea.

167 *reconceptualized mesmerism:* Eleanor Cummins, "How Hypnosis Works, According to Science," *Time*, April 28, 2022, https://time.com/6171844/how-hypnosis-works/; A. Gordon Hammer et al., "Hypnosis," *Encyclopedia Britannica,* September 6, 2025, https://www.britannica.com/science/hypnosis; Christy Martin, "Mesmerized: The Controversy Around Animal Magnetism," *Distillations Magazine*, December 6, 2011, https://www.sciencehistory.org/stories/magazine/mesmerized/.

167 *clinical hypnosis has clear benefits:* Kirsten Weir, "Uncovering the New Science of Clinical Hypnosis," *Monitor on Psychology*, April 1, 2024, https://www.apa.org/monitor/2024/04/science-of-hypnosis.

167 *empowering and even pain-alleviating effects:* Trevor Thompson et al., "The Effectiveness of Hypnosis for Pain Relief: A Systematic Review and Meta-Analysis of 85 Controlled Experimental Trials," *Neuroscience & Biobehavioral Reviews* 99 (April 2019): 298–310, https://doi.org/10.1016/j.neubiorev.2019.02.013.

167 *measurable effects on brain activity:* Cummins, "How Hypnosis Works."

167 *hypnosis changes activity:* Mathieu Landry et al., "Brain Correlates of Hypnosis: A Systematic Review and Meta-Analytic Exploration," *Neuroscience & Biobehavioral Reviews* 81 (October 2017): 75–98, https://doi.org/10.1016/j.neubiorev.2017.02.020.

167 *participants with nearsightedness:* C. Graham and H. W. Leibowitz, "The Effect of Suggestion on Visual Acuity," *International Journal of Clinical and Experimental Hypnosis* 20, no. 3 (1972): 169–86, https://doi.org/10.1080/00207147208409288.

168 *study in those with nearsightedness:* Eugene P. Sheehan et al., "A Signal Detection Study of the Effects of Suggested Improvement on the Monocular Visual Acuity of Myopes," *International Journal of Clinical and Experimental Hypnosis* 30, no. 2 (1982): 138–46, https://doi.org/10.1080/00207148208407379.

168 *potential alternative explanations:* Amir Raz et al., "Critique of Claims of Improved Visual Acuity After Hypnotic Suggestion," *Optometry and Vision Science* 81, no. 11 (2004): 872–79, https://doi.org/10.1097/01.OPX.0000145032.79975.58.

168 *how hypnosis may improve hearing:* Kenneth Sterling and James G. Miller,

"The Effect of Hypnosis upon Visual and Auditory Acuity," *American Journal of Psychology* 53, no. 2 (1940): 269, https://doi.org/10.2307/1417422.

170 *Candy discussed these points:* Brown, *Testing Prayer*, 194–233; Brown et al., "Study of the Therapeutic Effects."

Chapter 10: Miracles Happen

171 *her book,* Testing Prayer*:* Candy Gunther Brown, *Testing Prayer: Science and Healing* (Harvard University Press, 2012), 99–154.

173 *historical and legal standards for evaluating evidence:* Maksymilian Del Mar, "What Does History Matter to Legal Epistemology?," *Journal of the Philosophy of History* 5, no. 3 (2011): 383–405, https://doi.org/10.1163/187226311X599871.

174 *Marsha's case:* Clarissa Romez et al., "Case Report of Instantaneous Resolution of Juvenile Macular Degeneration Blindness After Proximal Intercessory Prayer," *Explore* 17, no. 1 (2021): 79–83, https://doi.org/10.1016/j.explore.2020.02.011.

176 *Marsha's eyesight had remained intact:* Romez et al., "Case Report of Instantaneous Resolution of Juvenile Macular Degeneration Blindness."

177 *a boy named Derek:* Clarissa Romez et al., "Case Report of Gastroparesis Healing: 16 Years of a Chronic Syndrome Resolved After Proximal Intercessory Prayer," *Complementary Therapies in Medicine* 43 (April 2019): 289–94, https://doi.org/10.1016/j.ctim.2019.03.004.

178 "*After the sermon*"*:* Romez et al., "Case Report of Gastroparesis Healing."

180 *Peter was born:* Clarissa Romez et al., "Case Report of 11 Years of Severe Malabsorption, Muscular Atrophy, Seizures, and Immunodeficiency Resolved After Proximal Intercessory Prayer," *Advances in Mind-Body Medicine* 28, no. 2 (2024): 40–55.

182 "*While attending the Healing Rooms*"*:* Romez et al., "Case Report of 11 Years of Severe Malabsorption."

Chapter 11: Believing in Miracles

188 *some neurons make direct connections:* David Linden, "Can a Neuroscientist Fight Cancer with Mere Thought?," *New York Times*, March 18, 2023, https://www.nytimes.com/2023/03/18/opinion/cancer-brain-mind-body.html.

188 *levetiracetam doesn't slow:* Jia-Shu Chen et al., "The Effect of Levetiracetam Treatment on Survival in Patients with Glioblastoma: A Systematic Review and Meta-Analysis," *Journal of Neuro-Oncology* 156, no. 2 (2022): 257–67, https://doi.org/10.1007/s11060-021-03940-2.

188 *Pain-sensing nerve cells:* Mohammad Balood et al., "Nociceptor Neurons Affect Cancer Immunosurveillance," *Nature* 611, no. 7935 (2022): 405–12, https://doi.org/10.1038/s41586-022-05374-w.

188 *Various nerve cells:* Varun Venkataramani et al., "Glutamatergic Synaptic Input to Glioma Cells Drives Brain Tumour Progression," *Nature* 573, no. 7775 (2019): 532–38, https://doi.org/10.1038/s41586-019-1564-x.

189 *which together indicated a tumor:* Attila Rácz et al., "Age at Epilepsy Onset in Patients with Focal Cortical Dysplasias, Gangliogliomas and Dysembryoplastic Neuroepithelial Tumours," *Seizure* 58 (May 2018): 82–89, https://doi.org/10.1016/j.seizure.2018.04.002.

189 *They cause death:* K. Lote et al., "Survival, Prognostic Factors, and Therapeutic Efficacy in Low-Grade Glioma: A Retrospective Study in 379 Patients," *Journal of Clinical Oncology* 15, no. 9 (1997): 3129–40, https://doi.org/10.1200/JCO.1997.15.9.3129; Timothy J. Brown et al., "Management of Low-Grade Glioma: A Systematic Review and Meta-Analysis," *Neuro-Oncology Practice* 6, no. 4 (2019): 249–58, https://doi.org/10.1093/nop/npy034.

190 *glioma in an older person:* Maximilian Scheer et al., "Spontaneous Remission of a 'Diffuse Glioma'—A Case Report," *Interdisciplinary Neurosurgery* 28 (2022): 101452, https://doi.org/10.1016/j.inat.2021.101452; Masahiro Ishihara et al., "Spontaneous Complete Regression of a Brain Stem Glioma Pathologically Diagnosed as a High-Grade Glioma," *Child's Nervous System* 33, no. 12 (2017): 2177–80, https://doi.org/10.1007/s00381-017-3570-3; Warren Matthew Rozen et al., "Spontaneous Regression of Low-Grade Gliomas in Pediatric Patients Without Neurofibromatosis," *Pediatric Neurosurgery* 44, no. 4 (2008): 324–28, https://doi.org/10.1159/000134925.

190 *Localized infections:* Annamaria Vezzani et al., "Infections, Inflammation and Epilepsy," *Acta Neuropathologica* 131, no. 2 (2016): 211–34, https://doi.org/10.1007/s00401-015-1481-5.

190 *mesial temporal sclerosis:* R. A. Prayson et al., "Mesial Temporal Sclerosis. A Clinicopathologic Study of 27 Patients, Including 5 with Coexistent Cortical Dysplasia," *Archives of Pathology & Laboratory Medicine* 120, no. 6 (1996): 532–36.

191 *nine times more likely:* R. A. Bronen et al., "Focal Cortical Dysplasia of Taylor, Balloon Cell Subtype: MR Differentiation from Low-Grade Tumors," *AJNR: American Journal of Neuroradiology* 18, no. 6 (1997): 1141–51; Luc Bauchet et al., "Epidemiological Analysis of Adult-Type Diffuse Lower-Grade Gliomas and Incidence and Prevalence Estimates of Diffuse IDH-Mutant Gliomas in France," *Neurochirurgie* 71, no. 3 (2025): 101627, https://doi.org/10.1016/j.neuchi.2024.101627; Javier A. López-Rivera et al., "Incidence and Prevalence of Major Epilepsy-Associated Brain Lesions," *Epilepsy & Behavior Reports* 18 (2022): 100527, https://doi.org/10.1016/j.ebr.2022.100527.

191 *10 percent of patients:* Susanne Fauser et al., "Clinical Characteristics in Focal Cortical Dysplasia: A Retrospective Evaluation in a Series of 120 Patients," *Brain* 129, no. 7 (2006): 1907–16, https://doi.org/10.1093/brain/awl133.

195 *"remarkable" designation:* Dirk J. Kruijthoff et al., "Prayer and Healing: A Study of 83 Healing Reports in the Netherlands," *Religions* 13, no. 11 (2022): 1056, https://doi.org/10.3390/rel13111056.

Afterword: The Mystery of Healing

197 *concept of manifestation:* James R. Doty, *Mind Magic: The Neuroscience of Manifestation and How It Changes Everything* (Avery, 2024).

INDEX

ABOUT THE AUTHOR

Joshua W. Brown, PhD, is a professor of psychological and brain sciences at Indiana University, where he directs the Cognitive Control Lab. He is the author of over ninety scientific publications, director of the Global Medical Research Institute, and executive producer of the Angel streaming series *Miracle*. His neuroscience research has been featured on NPR and the Discovery Channel, and his research on miracles has been featured in *The New York Times*.